W9-AYD-296

FRONTIERS

RECENT SCIENCE ESSAY
COLLECTIONS BY ISAAC ASIMOV

The Relativity of Wrong (1988)

Past, Present and Future (1987)

Far as Human Eye Could See (1987)

The Dangers of Intelligence (1986)

The Subatomic Monster (1985)

X Stands for Unknown (1984)

FRONTIERS

New Discoveries About Man and His Planet, Outer Space and the Universe

Isaac Asimov

TRUMAN TALLEY BOOKS
E. P. DUTTON
NEW YORK

Published in the United States by
Truman Talley Books • E. P. Dutton,
a division of Penguin Books USA Inc.,
2 Park Avenue, New York, N.Y. 10016.

Published simultaneously in Canada
by Fitzhenry and Whiteside, Limited, Toronto.

Library of Congress Cataloging-in-Publication Data
Asimov, Isaac, 1920–
Frontiers : new discoveries about man and his planet, outer space and the universe / Isaac Asimov. — 1st ed.
p. cm.
Expanded version of columns which originally appeared in the Los Angeles times
"Truman Talley books."
ISBN 0-525-24662-2
1. Science—Popular works. 2. Outer space—Exploration—Popular works. 3. Cosmology—Popular works. I. Title.
Q126.A75 1990
500—dc20 *89-39778*
CIP

Designed by Margo D. Barooshian

1 3 5 7 9 10 8 6 4 2

First Edition

All the essays in this book originally appeared in The Los Angeles Times.

to Read Evans,
the very pattern
of a loyal reader

CONTENTS

INTRODUCTION 1

I FRONTIERS OF EARLY MAN

Our Ancient Ancestors 7 ■ How Old Are We? 10 ■ On Your Hind Legs 13 ■ Hands at Last 16 ■ A Bone Speaks Volumes 19 ■ Man's First Language 22 ■ Man's First Discovery 25 ■ Fueling the Fires 28 ■ Our Cousin the Coelacanth 31 ■ The Relentless Population Rise 34

II FRONTIERS OF SCIENCE

The Brightest X Rays 39 ■ Reward Delayed 42 ■ The Noblest Element 45 ■ Raising the Temperature 48 ■ Atom Smasher Supreme 51 ■ The Two Nothings 54 ■ Supercritical 57 ■ A Question of Priorities 60 ■ Suddenly, Thallium Plays a Role 62 ■ Breaking the Bond 66 ■ A Dream Comes True 69 ■ Getting Old 72 ■ The Elusive Quark 75

■ How Many Particles? 78 ■ Taming Antimatter 81
■ Improving on the Diamond 84
■ Cold, Cold Fusion 87
■ Tritium—Why It's Crucial 90
■ Forever Gone 93 ■ The Simpler Shape 96
■ The Dangerous Microorganism 99
■ The Telltale Flash 102 ■ The Genome Project 105
■ First Look at the DNA Molecule 108
■ The Head of a Pin 111 ■ Our Biological Clock 114

III FRONTIERS OF EARTH

The Shifting Earth 119 ■ The Wobbling Earth 122
■ Those Oceanic Hot Spots 125
■ The Big Crack 128 ■ The Central Heat 131
■ The First Cell 133
■ The Conquest of the Land 136
■ How Green Plants Began 139
■ Dinosaurs Everywhere 142 ■ Squashed Sand 145
■ Death of the Dinosaurs: A New Clue 148
■ Fossil Fact or Fiction 151
■ More Evidence of Feathered Fliers 154
■ The Biggest Flyer 157 ■ Past Monsters 160
■ The Most Successful Life-Form 163
■ The Homing Turtles 166
■ The Oddest Mammal 169 ■ Old Water 172
■ Lightning and Life 175 ■ Overkill 178
■ The Ozone Hole 181 ■ The Last Clean Place 184
■ Wetter and Warmer 187 ■ Leap Second 191
■ A Map Too Good to Be True 193
■ The Mislaid Island 196
■ When the Earth Was Too Hot and Too Cold 199
■ Ice Ages and the Plateau Effect 202
■ Moon Mysteries, Earth's History 205

IV FRONTIERS OF SPACE

The Cracked Crust 211 ■ Blast over Siberia 214 ■ Halley's Comet 217 ■ More About Halley's Comet 220 ■ The Largest Molecule 223 ■ Our Identical Twin 226 ■ Microwaves as Cloud-Zappers 229 ■ Space Rocks 232 ■ Close Call with an Asteroid 235 ■ Diamonds from Space 238 ■ The Dead World 241 ■ The Slow Breakdown 244 ■ Old Reliable 246 ■ Going Where the Energy Is 249 ■ An Ocean of Gasoline 252 ■ The Elusive Tenth 255 ■ The Little Rocket that Could 258 ■ The Lopsided Satellite 260 ■ Beware the Flare! 264 ■ Skimming the Sun 267 ■ The Invisible Cloud 270 ■ What's in a Name? 273 ■ Pluto and Charon: The Dumbbell Worlds 276 ■ The Case of the Missing "Planet" 279 ■ The Falling Moon of Mars 282 ■ Life on Mars Revisited 285 ■ A Little Brighter 288 ■ Space Pollution 291 ■ Where Do We Go from Here? 293 ■ The End? 296 ■ Are We Alone? 299

V FRONTIERS OF THE UNIVERSE

The Supernova Next Door 305 ■ The Planet Hunt 308 ■ Far Beyond 311 ■ The Giveaway Bursts 313 ■ The Neutron Surprise 316 ■ The Invisible Dust Clouds 319 ■ The Weakest Wave 322 ■ The Relativity Test 325 ■ Neutrinos from Afar 328

■ The White Dwarf Clock 331
■ The Gamma Giveaway 334
■ The Force that Can Swallow a Star 337
■ The Cluster Yardstick 340
■ Gravity Plays Tricks 343
■ The In-Between Objects and the Missing Mass 346
■ Galaxies in Collision 349
■ Ten Billion Light-Years Away 352
■ Seeing the Past 355 ■ The Fastest Telescope 358
■ The Oldest Birthday 361 ■ Superstars? 363
■ The Baby Pulsar 367 ■ Beyond the Beyond 370
■ Why Are Things as They Are? 373
■ Where the Universe Ends 376

INDEX 379

Introduction

Since men first learned to think analytically and use available materials to make life easier and more secure, we have been faced with the crossing of an endless succession of frontiers. In the sciences, these boundaries have always been ports of embarkation for the unknown, and careful reviews have necessarily preceded understanding and progress.

A similar process of review and reevaluation is important to any writer who sets out to explain the complexities of modern science to those not formally trained in the discipline. This maxim has been reinforced for me since 1986, when I began to write a weekly science column for The Los Angeles Times Syndicate. The experience has been great fun and it has prompted my desire to expand that work into book form. *Frontiers*, then, is a collection of broad-based observations that reviews recent advances in science and reexamines and evaluates the historic achievements that led to the focusing of our knowledge. My hope is that the informal approach of this collection will serve to open new vistas for those who share my wonder at the awesome frontiers that face us as we move toward a better understanding of our seemingly endless universe.

In preparing this book, it has occurred to me that its contents appear to have a rather idiosyncratic character. I tend to follow my interests and some things simply interest me more than others. For this reason, the discussion will be centered more on physics and astronomy than on the medical sciences, for example, which already receive more newspaper and magazine coverage than all other branches of science put together.

As the book deals with the frontiers of science, the conclusions of the various essays are sometimes tentative. After all, science *is* tentative: it always stands ready to extend or correct itself. Thus, I mention a recent finding on the concentration of oxygen in ancient atmospheres and mention my own doubts about the value of the result. Or I examine an apparently exciting discovery of arcs of light in the sky and then point out that these arcs seem to have turned out to be a kind of optical illusion.

The results scientists achieve on some frontier subject are quite likely to be contradictory, but that is characteristic of a frontier. For instance, in the numerous essays in the book, I take up the matter of the age of the universe. In some cases, new findings lead to the thought that the universe is some 10 billion years old, and in other cases, an age of 20 billion years is indicated. Well, which is it? The answer is that we can't say for certain. It is a difficult point to determine and different lines of research point in somewhat different directions. This is not a flaw in science but is rather one of the glories: there is free argument on controversial points and there are multiple paths to the goal, where some of the paths may prove dead ends. The time will surely come when the matter of the age of the universe will be settled, as the age of the Earth has been. Until then, the reader may well be interested in knowing the different ways of tackling the problem and weighing for himself the relative values of the various arguments and results.

Finally, the reader is bound to find considerable overlap among some of the essays. After all, each has been written to stand on its own feet. Therefore, two essays on similar subjects may require something of the same background, and I can only plead for understanding and forgiveness.

Despite all these caveats, however, I am very hopeful that this book will give the reader a feel for what many scientists are discovering *at the present time.* Science is very much a living subject, and it has never been more so than it is now. More scientists today, with more techniques at their immediate disposal, are investigating more subjects, with more enthusiasm than ever before. As a result, the fields of human knowledge are expanding more vigorously and astonishingly than ever before.

Isaac Asimov

October 1, 1989

I

FRONTIERS OF EARLY MAN

Our Ancient Ancestors

The human species is a newcomer on Earth. We haven't been here for long compared to Earth's mighty lifetime, but we have been around longer than we used to think. And scientists are still periodically surprising themselves with new measurements that show us, or our forebears, to be more and more ancient.

Until modern times, Western scholars, even scientists, took it for granted that humanity (and Earth itself) was only six thousand years old or so, because that was what the Bible seemed to imply. As early as 1797, however, an Englishman, John Frere, discovered crudely fashioned flint tools that must have been made by primitive human beings. These tools were discovered thirteen feet underground. Objects that remain undisturbed are slowly covered by dust and mud that turns into rock, and any tools buried that deep must be far more than six thousand years old.

Later, a Frenchman named Édouard Lartet found an ancient mammoth tooth on which was scratched an excellent drawing of a mammoth. It could only have been drawn by a human being who lived at the same time as the long-extinct mammoths did.

In time, bony remains were found of organisms that were not quite human beings but were closer to human beings in skeletal structure than they were to apes. These were called *hominids*, and they represented a long line of organisms that were ancestors (or collateral branches) of modern human beings: *Homo sapiens*.

Hominids were known to be ancient, but how ancient?

Scientists could only guess rather roughly by estimations from the depths at which the remains were found, and from the kind of bones of other animals that surrounded them. It was suspected that hominids might have existed on Earth for hundreds of thousands of years, but the dating was not firm.

In 1896, however, radioactivity was discovered. It was found that certain kinds of atoms were unstable and broke down at a fixed rate, which could be measured. Thus, uranium broke down to lead at a rate at which half the uranium would be turned into lead in 4.6 billion years. In 1907, an American, Bertram B. Boltwood, suggested that rocks that contained uranium would also have to contain lead. From the proportions of uranium and lead, you could calculate how much uranium had decayed and therefore how old the rock must be.

This was the beginning of "radioactive dating," which could be used to determine the age of undisturbed rocks. Some of the rocks discovered have, by radioactive dating, proved to have lain undisturbed for about 3.5 billion years, so the Earth must be older than that. Meteorites, which have been undisturbed from the very beginning, show ages of about 4.6 billion years, which is now taken as the age of the Earth—and the solar system.

Naturally, if you discover hominid bones embedded in rock, and you determine the age of the rock, you will also have determined the age of the bones. Not all rocks have enough uranium in them to allow a determination, but they all have some of the common element potassium in them. Certain potassium atoms are radioactive and break down to the inert gas argon at such a rate that half the potassium is gone in 1.3 billion years. By measuring the potassium and the trapped argon bubbles in rocks, the length of time since those rocks formed and those bones were encased can be determined.

Naturally, as time passed, techniques for such dating improved, and usually hominids turned out to be older than previously thought. In September 1987, scientists from the University of Utah dated rocks in Kenya that contained ancient tools. The rocks had been thought to be 500,000 years old, but the new measurements indicate they are at least 700,000 and possibly 900,000 years old.

And there are hominids older than this, too (they seem to have evolved in eastern and southern Africa; this isn't surprising because that is where our closest nonhominid relatives, chimpanzees and gorillas, live). In Olduvai Gorge in East Africa, hominid skulls and primitive tools were discovered in rocks that surprised scientists by proving to be about 1.8 million years old. The hominids were part of our genus, *Homo*, and are referred to as *Homo habilis*.

Before *Homo habilis* there existed even more primitive hominids that were too different from us to be considered part of *Homo* but that were still hominids. For instance, they had hips and legs just like ours and could walk erect as easily as we do. The oldest of these bears the name *Australopithecus afarensis*. Fossilized bones that seem to place it at about 4 million years have been discovered.

Undoubtedly there must be even earlier specimens. It seems reasonable to suppose that there have been hominids on Earth for 5 million years. This would make the human species and its hominid forebears some eight hundred times as old as scholars thought as recently as two centuries ago. But (just to keep a sense of proportion) hominids have still existed only in the last one-thousandth of Earth's existence.

How Old Are We?

How old are we? By *we* I mean the group of organisms known variously as "present-day man," "modern man," "human beings," or *Homo sapiens sapiens*. The answer may be, it now turns out, more than twice as old as we thought we were.

To see what this means, let's go back to 1856. In the west German valley of the Neander River (*Neanderthal* in German), workmen clearing out a limestone cave came across some bones. There was nothing unusual about this. The common practice was to throw away such bones. And so these were, but word of it got to a professor at a nearby school. He managed to get to the site and salvage about fourteen of the bones, including a skull.

The bones were clearly human, but the skull in particular showed some interesting differences from that of an ordinary human being. It had pronounced bony ridges over the eyes. It also had a backward-sloping forehead, a receding chin, and unusually prominent teeth. The remains were quickly dubbed "Neanderthal man," and a loud controversy immediately arose. Were they the remains of an ancient and primitive ancestor of modern human beings, or were they from an ordinary human being who had had some sort of bone disorder?

Later, other bony remains that included just such skulls were found elsewhere in Europe and in the Middle East. There couldn't be so many people with the same bone disease. It therefore came to be accepted that Neanderthal man was an early, somewhat primitive type of human being. Anthropologists began to refer to Neanderthal man as *Homo neanderthalensis* (modern man was called *Homo sapiens*, *sapiens* meaning "sapient," or "knowing"). Both belonged to the genus *Homo*.

Eventually though, the differences between Neanderthal man and modern man seemed so small that anthropologists began to think of them as members of two subspecies. Neanderthal man was called *Homo sapiens neanderthalensis* and modern man *Homo sapiens sapiens*.

Neanderthal man may have originated from earlier, still more primitive ancestors as long as 250,000 years ago. At some time and some place, some Neanderthals underwent the minor evolutionary changes required to gain modern attributes. We don't know exactly when or exactly where, because the Neanderthals were few in number and were too smart to get themselves trapped under conditions in which they would fossilize, so we have very few fossils to judge from.

Still, old skeletons that are exactly like modern skeletons have been found, and, judging from them, modern man must have developed at least 40,000 years ago. This may have happened in northern Africa, though this is very uncertain.

The latest Neanderthal skeletons are about 35,000 years old. For a while, then, modern man and Neanderthal man were both on the Earth together (in Europe, chiefly, for that is where most Neanderthal fossils have been found). So it may be that Neanderthal man and modern man lived together for only five thousand years before Neanderthal man was gone.

Presumably, when the two subspecies encountered each other they competed for food and habitats, and Neanderthal man lost out. Why? It's uncertain. There is some ground for thinking that individual Neanderthal men were stockier and stronger than individual modern men. However, they may have been less agile.

Or the moderns may have been more ingenious. My own favorite theory is that modern man invented long-distance weapons, such as slings or bows and arrows, with which to attack the Neanderthals from a distance and thus avoid the dangers of close combat. The poor Neanderthals may therefore have lost almost every battle and shrunk steadily in numbers until the growing numbers of *Homo sapiens sapiens* were left undisputed overlords of the Earth.

But a report published in February 1988 by a group of French and Israeli anthropologists raises new questions about early man's relationship to the Neanderthals. The report details the findings in an Israeli cave of the skeletal remains of some thirty human beings that seem to be those of *Homo sapiens sapiens.* Stone tools found with those remains were tested for age by a technique called *thermoluminescence* (the production of light on heating), and if the results are correct, the skeletons are about 90,000 years old.

If so, this would mean that modern man split off from the Neanderthal stock more than twice as long ago as had been thought, and there was that much more time for the development of differences that perhaps do not show up in the bones. Perhaps, if the results hold up, anthropologists may decide to consider Neanderthal man and modern man as different species again.

Also, if Neanderthal man and modern man coexisted on Earth not for 5,000 years, but for 55,000 years, why should modern man have taken that long to wipe out Neanderthal man? Were the Neanderthals smarter than we thought? Did they put up a much better fight?

Scientists must now wrestle with these questions. But it's a sad thought that modern man, if he really tried, could nowadays probably wipe himself out in 55,000 seconds.

On Your Hind Legs

The oldest "hominids," creatures who were more humanlike than apelike, were the *australopithecines*. This is a misnomer, for the word is from the Greek and means "southern apes." The australopithecines are southern, all right, for their fossil remains were first uncovered in the Southern Hemisphere (South Africa, to be exact), but they were *not* apes.

They may have had the size and build of rather small apes, and they may have had brains no larger than those of chimpanzees, but they walked upright. They had feet and hips and a spine, just like ours, and must have walked just as erect and easily as we do.

This trick of walking on your hind legs is the very oldest of human characteristics. The australopithecines had evolved and were upright walkers as much as 4 million years ago. No one else did it. Chimpanzees and gorillas have feet with opposable thumbs so that, in effect, they have four hands. They raise themselves to their hind legs only rarely and uncomfortably. They don't have our feet, which do not have opposable thumbs and are specialized for walking only. They don't have our S-shaped spine and our hipbone arrangement, which make it easy for us to remain upright for extended periods.

But why did the australopithecines develop upright posture? What good did it do them? What survival value did it have? One possibility is that it gave them added height from which to sight food, or danger, from afar. For that, however, one needed to rise to one's hind legs only occasionally and temporarily.

A romantic view is that rising to one's hind legs freed one's arms. It made it possible to develop hands that were primarily used for the manipulation of the environment, for investigation, for making tools. All this put a premium on better eyes and brains, so that our brains grew, and we became fully human.

This was undoubtedly so as an *eventual* side effect, but not an immediate one. After the first australopithecines began to walk erect, they continued to exist for 2 million years before their descendants developed a brain large enough to enable them to make stone tools and to develop the first signs of an intellect we might consider human.

But then, what good did walking erect do the australopithecines during the 2 million years in which they remained small-brained and did not use their hands for making tools?

Mary Leakey (who has been perhaps the most famous of the finders of hominid remains, along with her late husband, Louis, and her son, Richard) and her collaborators have a suggestion: The australopithecines, she thinks, were scavengers. They were not large enough to kill the great herbivores of Africa, or smart enough to organize hunting parties. Instead, they would seize on the remains of those large animals that were killed by such predators as lions and leopards. In short, they had the life-style of jackals, hyenas, and vultures (which is rather embarrassing to think of now).

If the australopithecines had waited for a kill to be made in their vicinity, they would have had to wait a long time. Most scavengers have to do that, since the necessity of caring for their young often pins them close to their dens.

However, by the development of the ability to walk erect, the australopithecines' arms were indeed freed, not for making tools, but for carrying their infants.

We have a picture, then, of these rather small hominids, holding their young in their forelimbs and scurrying along on their hind limbs, following the herds of wildebeests and zebras and waiting for the kills in which they might eventually share.

As it happens, my wife, Janet (who is a psychiatrist), has long speculated on some of the aspects of walking upright and for some years she has been convinced that it was carrying the young that was important.

She has suggested to me that, because of human hairlessness, infants could not hold on to the mother's body hair and therefore had to be cradled in her arms. Of course, we don't know when, in the course of human evolution, body hair was lost. We don't know whether the australopithecines were as hairy as apes or as hairless as we, or in between. If, however, the body hair began to disappear (especially in females) at about the time they were rising to their hind feet, that would have given an added impulse to the need for child carrying.

My wife also points out that infants may have been quieter when held in the left arm, nearer the soothing sound of the heartbeat (to which they were accustomed in the uterus); this left the right arm free for the manipulation of the environment and may have resulted in human beings' becoming right-handed (as 90 percent of us are). Our cousins, the apes, after all, show no signs of handedness, using right and left limbs with equal ease.

Hands at Last

The fairly recent discovery of a few little bones has suddenly raised interesting questions about toolmaking among early creatures who more closely resembled human beings than apes.

These earliest hominids were *australopithecines* ("southern apes") because their skeletons were first found in South Africa, and they seem to have been confined to southern and eastern Africa. They were not apes, however, for the bones of their legs and hips are very much like ours and they walked erect just as easily and well as we do.

The very first australopithecines may have evolved as long as 5 million years ago. The very last australopithecines died out perhaps 1 million years ago. This means they lasted for 4 million years and can therefore be considered a successful group of beings.

The earliest australopithecines were small creatures, less than four feet tall and weighing only sixty-five pounds or so. Their brains were no larger than those of chimpanzees, but they walked erect and were probably brighter than chimpanzees.

As the centuries passed, the australopithecines evolved and developed into several species. Modern scientists, sorting out the fossilized bones of these creatures, have identified at least four of these species. In general, as time went on, australopithecines and their brains grew larger.

Some 2.5 million years ago, the species *Australopithecus robustus* made its appearance. It may have been up to five feet tall and weighed as much as 110 pounds. Its brain was about one-third the size of ours, a bit larger than that of a gorilla. The very largest australopithecines may have been as large as we are.

A mere increase in size, however, does not alone suffice to make the australopithecines more human. About 2 million years ago, some form of australopithecines (we are not sure which) developed skulls that were closer to the modern human skull than any australopithecine had formerly possessed. The new creature so closely resembled us that it was put into our genus and given the proud name *Homo* (Latin for "man").

The earliest example of genus *Homo* that we know of is *Homo habilis*, a rather small creature, certainly smaller than the larger australopithecines. From *Homo habilis* there descended the larger and brainier *Homo erectus*, who were the first hominids to wander out of Africa and into Asia. And from *Homo erectus* eventually came *Homo sapiens*, first a variety called "Neanderthal man" and then us, "modern man."

Habilis is from a Latin word meaning "handy" or "skillful." *Homo habilis* therefore is "handy man." It is called that because where its fossilized bones are found there are also found small stone objects that look like tools. Such tools were not found in the neighborhood of australopithecine fossils.

It seemed, therefore, that only creatures who were members of genus *Homo* were bright enough and ingenious enough (skillful and "handy" enough) to form and use stone tools. Australopithecines, although they walked like men, were just too limited in brain organization (even when their brains were as large as those of *Homo habilis*) to handle stone. Or perhaps the australopithecines didn't have hands that were sufficiently flexible and efficient to handle stone.

That is one of the frustrations of trying to study hominid fossils. We don't have many of them to begin with, but what we do have consists of skulls, teeth, hipbones, and thighbones for the most part. We don't find bones of the hand, and it is the hand that, next to the skull, is most characteristic of humanity.

But now, in a cave in South Africa, fossil remnants of *Australopithecus robustus* have been found that include bones of the hand. Hands at last.

It turns out that the larger australopithecines had fingers, and thumbs, too, that were shaped like ours. The australopithecines, at least the later and larger ones, had thoroughly human hands.

It seems only reasonable to suppose that the bigger australopithecines, with human hands and with brains at least as large as that of *Homo habilis*, were skillful enough to handle tools.

But there are tools and tools. It is quite likely that australopithecines might have used a tree limb or a thighbone as a club. They may perhaps even have shaped wood and bone into fancier tools. However, wood and bone don't last as stone does, so we don't find wood and bone relics that are millions of years old. And it takes far more patience and skill to make the more useful stone tools, and there the australopithecines may have fallen short.

To discover whether australopithecines used stone tools, we would have to find stone tools associated with australopithecine relics. Until now any stone tools that have been found were assumed to have been made by some creature of genus *Homo*. The latest discovery indicates that more careful searching and fewer preconceived notions may uncover creatures other than genus *Homo* who also handled stone implements.

A Bone Speaks Volumes

The ability to speak, to make a variety of complicated sounds so rapidly and clearly as to communicate information and abstract ideas, is a peculiarly human trait. We can do it, but could any other creatures do it before "modern man" (*Homo sapiens sapiens*) appeared on Earth, at least 50,000 years ago? Many anthropologists have answered that question with a no, but a startling new finding makes it seem the answer might be yes.

Even our closest living relatives, the great apes, do not speak and, what is much more important, cannot speak. An ape's larynx, or voice box, and the region around it are not so arranged that they can make the rapid variety of sounds we can. Chimpanzees and gorillas have been taught to communicate simple ideas, but only by signs and gestures. Even the brightest of them, and the most assiduously trained of them, can no more speak than they can fly. They are not anatomically equipped to do either.

The same is true of other animals. Certain birds, such as parrots and mynahs, with totally different throat anatomies from ours, can nevertheless be trained to mimic the sounds we make, but, of course, without understanding. Dolphins can easily make an even greater variety of sounds than we can, but we cannot tell whether they use them in speech in something like the human manner.

This, however, leaves the question as to when human ancestors first began to speak. Did hominids more primitive than "modern man" manage to do so?

It seems unlikely that the truly primitive hominids—the australopithecines, *Homo habilis* or *Homo erectus*—all of whom lived between 5 million and 200,000 years ago, could speak. Their brains simply weren't large enough.

This leaves Neanderthal man, who first appeared on Earth perhaps as much as 300,000 years ago and didn't become extinct till 30,000 years ago. His skeletal remains show him to have been so similar to us that he is often considered a subspecies of "modern man" and is called *Homo sapiens neanderthalensis*. His brain was as large as ours or even a trifle larger, but there was more of it to the rear of the skull and less in the front, whatever that might mean.

The question, then, is whether Neanderthal man could speak. The key to the answer lies in the hyoid bone. This is a small U-shaped bone set at the base of the tongue (*hyoid* is from the Greek for "U-shaped"). It is not connected to any other bone but is connected to the larynx by eleven small muscles in two groups. These can lift and depress the larynx and make it possible to sound various vowels and consonants in rapid succession. Without the hyoid bone, we wouldn't be able to move the larynx in this way, and so we might call the hyoid the "speaking bone."

The skeletal remains of Neanderthal men have till now never included a hyoid bone, so the conclusion that they couldn't speak seemed reasonable. They may have communicated fairly well, but only by sign language and grunting.

However, such evidence really isn't conclusive. The hyoid bone is quite small, only an inch wide plus two narrow horns, and is unconnected to other bones. When a body decays, the hyoid bone falls loose, and it may be deposited far away from the rest of the skeleton. In the remains that have been found, no remnants of the larynx are present either, for that matter; so we can't really be sure Neanderthals couldn't speak.

But then, in April 1989, a startling find was reported by

Baruch Arensburg of Tel Aviv University and co-workers from the University of Bordeaux in France and Moorhead State University in Minnesota. In a cave in Mount Carmel in Israel, they found the remains of Neanderthal men, including a hyoid bone just about exactly the shape and size of the one "modern man" possesses. The age of the bone is estimated at 60,000 years.

The implication is that Neanderthal man was anatomically capable of speech. He coexisted with modern man for perhaps as much as 50,000 years, and he could have learned speech from these advanced cousins of his.

I must admit that I have a personal interest in this. Science-fiction writers often deal with prehistoric times and with Neanderthal man in particular. Back in 1939, a good friend of mine, Lester del Rey, wrote a very touching story titled "The Day Is Done" about the last Neanderthal man, who was humanely cared for by the surrounding "modern men" but who died in despair because of his sense of inferiority. For one thing, he couldn't speak.

I never accepted that. I felt Neanderthal man was too like us to be unable to speak, and in 1958, I wrote a story titled "The Ugly Little Boy," in which a Neanderthal child is brought into the present and learns to speak English as well as we do.

Lester and I have argued repeatedly over this question. I shall be sure to break the news of the hyoid bone's discovery to him gently.

Man's First Language

When the Cro-Magnon men painted their colorful animals deep in the caves of present-day France and Spain 25,000 years ago, what language did they speak? Would you believe that there are scientists who are seriously trying to answer that question?

How can one possibly find out? Ancient people may leave behind their bones and their tools and even their art, but they don't leave any record of their language. They'd have to be able to write to do that, and writing was invented only about 5,000 years ago.

In a way, though, they do leave records of their languages, because languages aren't completely independent of one another. There are similarities, for instance, among such languages as Portuguese, Spanish, Catalan, Provençal, French, Italian, and (believe it or not) Romanian. These are all called Romance languages, because they are all similar, not only to each other, but to the old Roman language we call Latin.

This is not a mystery. Latin was once the common language of Western Europe in the days of the Roman Empire. After the fall of that empire and the temporary decline of education and other aspects of civilization, the Latin dialects in different parts of what had been the empire drifted apart and eventually developed into new languages. You can still detect similarities in vocabulary and grammar, however.

Suppose, then, that you had only these Romance languages, but that Latin had died out so completely that we had no record of it whatever. Might it not be possible, then, to go through the various Romance languages, study all the similarities, and construct a common language from which all might have developed? And if one did, might that constructed language not be something like Latin?

If you want to go even further back, there are similarities between Latin and Greek. The ancient Romans recognized this and adopted the more sophisticated grammatical principles that had been used in Greek and applied them to their own language. Must there not have been, then, an older language from which Greek and Latin both developed?

The surprising answer to this question came when the British began to seize control of India in the 1700s. The prime purpose was to engage in trade that would enrich Great Britain, but there naturally were scholars among the British who were interested in Indian civilization for its own sake. Among these was Sir William Jones, who studied an old Indian language, Sanskrit, which, like Latin, was no longer in use but had given rise to later variations.

Sanskrit survived in ancient epics and religious writings, however, and as Jones studied it, he found similarities in its vocabulary and grammar to both Greek and Latin. Furthermore, and this was the great surprise, there were similarities to the old Teutonic languages, such as Gothic, Old High German, and Old Norse. He even found similarities to Persian and to the Celtic languages.

In 1786, he concluded, therefore, that there was an "Indo-European" family of languages that stretched from Ireland to India and that probably stemmed from a single source. We might imagine that about 7000 B.C. there was an "Indo-European tribe" that lived, perhaps, in what is now Turkey. It spread outward in all directions, carrying with it its language, which evolved in different places as groups became isolated from each other. By studying all the similarities, might it not be possible to work up a kind of common language, a "Proto-Indo-European" that might indeed resemble what the original tribe spoke in 7000 B.C.?

This is made the more possible because in the 1800s the rules for the manner in which language changed with time were worked out by, among others, the Grimm brothers, who are better known today for the fairy tales they collected.

There are other language families that are not Indo-European. There is the Semitic group, which includes Arabic, Hebrew, Aramaic, and Assyrian. There is the Hamitic group, which includes certain early languages of Egypt, Ethiopia, and North Africa. There is the Ural-Altaic group, which includes Turkish, Hungarian, and Finnish (so that if Turkey was the original home of the Indo-Europeans, the vicissitudes of history have arranged to have a non-Indo-European language spoken there today).

Then there are the variety of languages spoken by Native Americans; by black Africans; by the Chinese and other peoples of the Far East; by the Polynesians; by the Australian Aborigines, and so on.

There are even languages that have no known connections to any other, such as ancient Sumerian and modern Basque.

If all of these were studied, might it be possible to work up an original language from which all descended? It would be an enormous task, but to linguists it would be a fascinating job.

The topic was discussed at a 1989 conference of historical linguists by Vitaly Schevoroshkia of the University of Michigan, who has been doing research on the subject.

It would be a useful endeavor, too, for if you could work out how human language evolved, it is possible you would work out, at the same time, the migrations and wanderings of early *Homo sapiens*.

Man's First Discovery

An early important discovery made by human beings or the more primitive hominids who were our ancestors was the use of fire, but we have never known exactly how long ago this discovery was made. But two South African archaeologists have reported evidence that the discovery may have come about much longer ago than had been thought.

Note that this is not a matter of the discovery of fire itself. Fire was a common event once forests came into existence about 400 million years ago. They could burn and, sparked by lightning, they did burn, so that animals were aware of fire, from which they fled, hundreds of millions of years before human beings appeared on the scene.

Human beings or prehuman beings, however, were the first organisms who did not merely run from fire. They tamed and used it. They carried a burning branch gingerly to some convenient place, sheltered it, fed the fire more fuel, and kept it burning.

At first, human beings or their ancestors had to depend on lightning to start a fire. If the fire went out, they had to borrow some from a neighboring tribe or wait for another lightning strike. It took many thousands of years before human beings learned to start fires—to be their own lightning, so to speak. We don't know exactly when that happened, either, or how.

However, just the use of fire, even by people who couldn't start their own, made all the difference. With fire, people could have light at night and warmth in winter. Fire made more hours of the day available for activity and enabled human beings to extend their range beyond the tropics into cooler regions. Fire frightened other animals, including the great predators, so that human beings could sleep safely in a cave that had a campfire at the mouth. This added to human security.

Fire could roast meat and make it taste better and also soften its fibers so that it was easier to chew. It could roast grain and make it soft and edible, greatly increasing the food supply. Fire also killed germs and parasites in food, reducing sickness.

Finally, human beings eventually learned to harden clay by baking it, thus making pottery possible. They also learned to heat sand and make glass out of it and to heat ore and make metal out of it. In short, fire was the indispensable beginning of human technology. This is one reason that no matter how intelligent dolphins and other sea creatures may be, without fire they can never develop even the crudest technology. And fires can't be built in a world of water.

When was fire first used? Until the 1980s the oldest traces of fire were detected in caves at Zhoukoudian near Beijing, the capital of China. There, traces of ancient campfires about 500,000 years old were found.

No human beings of the modern type *Homo sapiens* lived in those caves. In fact, *Homo sapiens* had not yet evolved. Living in those caves was a simpler hominid called *Homo erectus*, who looked more like us than like any ape but had a brain only a little more than half the size of ours.

Nevertheless, he was brainy enough to discover how to maintain and use fire, and for this we must be grateful to this ancestor of ours. But was this really the oldest apparent use of fire?

Perhaps not, for on December 1, 1988, two archaeologists, C. K. Brain and A. Sillen, reported finding evidence of much older campfires in certain caves in South Africa, about thirty-five miles west of Pretoria.

In these caves were found the remains of bones that seem to have been burned. Fresh bones are marrow-filled and fatty. If they are burned using an ordinary wood fire, the bones will burn brightly and hotly, just like a resinous torch. This apparently must have been what the primitive inhabitants did: used bone torches to light their way in the caves and to keep things warmer when it was cold.

Those burned bones are up to 1,500,000 years old, three times as old as the campfires of Zhoukoudian. There were no signs of charred bones buried in older layers in the caves, but once they began to appear, they continued to appear in more recent layers. Once fire was used, in other words, it continued to be used. It was too useful to be allowed to sink back into oblivion.

Dwelling in those ancient caves were older specimens of *Homo erectus*, so it would seem that those hominids began to use fire rather soon after they had evolved.

In fact, there are signs that also living in those caves at a different time was a still older and more primitive hominid, called *Australopithecus robustus*. This species became extinct not long after the time the fires in the cave were apparently being used, leaving the Earth to be dominated by *Homo erectus* and its descendants, *Homo sapiens*. Did *Australopithecus robustus*, before it died out, bequeath to us the use of fire? Not likely, in my opinion, but possible.

Fueling the Fires

Humanity's first source of energy beyond his own muscles was fire, the burning of fuel that could be easily located in the world about him. We may progress to other sources of energy, but there will always remain a demand for something simple, such as the burning of fuel, and scientists continue to look for better, more convenient fuels that will not be readily depleted.

Almost all fuels contain carbon or hydrogen atoms or both. Carbon and hydrogen combine with oxygen to produce light and heat, and all three types of atom are very common in the environment.

The first fuels used by human beings were wood and, to a much lesser degree, fats and oils from plants and animals. Wood, fats, and oils, which all contain carbon and hydrogen atoms, are renewable fuels because living things multiply and grow and produce more wood, fats, and oils to replace what is burned.

But not quite. As human numbers grew and human technology advanced, more and more fuel was needed and on the whole more fuel was burned or used for other purposes than was produced. The forests shrank.

Indeed, humanity could not have experienced the Industrial Revolution, during which the need for fuel was multiplied many times, had new fuels—chiefly coal, oil, and natural gas—not come into use. Coal is the residue of wood that was produced hundreds of millions of years ago and consists chiefly of carbon plus some hydrogen. Oil and gas are the residue of microscopic creatures that lived hundreds of millions of years ago and consist of carbon and hydrogen atoms.

We are now using enormous quantities of these "fossil fuels" (so-called because they are the remnants of ancient life), and new material of the sort is being produced at only a tiny rate. In effect, we are living on our capital, and eventually our reserves of coal, oil, and gas will be spent. Nor can we then return to wood, for at our present rate of use the forests (which are continuing to shrink as it is) will quickly be wiped out.

In addition, the fuels we use are dangerous, even while plenty still exist. Both coal and oil contain minor quantities of nitrogen and sulfur atoms, which produce oxides that are poisonous and acidic when they are burned. The atmosphere is polluted, and respiratory diseases increase. Acid rain is produced, helping kill forests and lakes.

Even the carbon atoms are dangerous, for they produce carbon dioxide when they are burned, so that our atmosphere is very slowly increasing its carbon dioxide content. That content is still very small, but carbon dioxide is an efficient retainer of heat, and even a small increase in atmospheric content may change the Earth's climate very much for the worse.

It is for these reasons—that the fuels of the Earth are limited in amount and are dangerous even while they last—that scientists are looking for alternative sources of energy.

But while I have mentioned the dangers inherent in burning nitrogen, sulfur, and carbon, what about hydrogen? Hydrogen burns more easily than any of these other elements and produces considerably more energy per pound when burned. What's more, hydrogen, on being burned, produces only water, which is harmless.

To be sure, hydrogen burns so easily as to have a tendency to explode, but so do gasoline and natural gas. We just have to be careful in the way we handle them.

The real problem is that hydrogen does not occur as such in nature. It can't just be chopped down, as wood can, or dug for like coal, or drilled for like oil. It must be produced chemically from substances that contain hydrogen atoms.

For instance, coal, oil, and gas all contain hydrogen atoms, and you can obtain pure hydrogen from them. But it takes some energy to do this. You have to burn some oil to get hydrogen out of other oil and, in this way, you end up with less fuel than you started with.

Scientists are looking for ways to produce hydrogen from fuels without using energy. Such non-energy-consuming reactions usually require catalysts, and finding the right catalyst isn't easy. Besides, when all the fuels are gone, then, catalysts or not, there will be no way of obtaining hydrogen.

Is there anything that contains hydrogen that is not a fuel? Yes, there is water, the weight of which is one-eighth hydrogen. The only trouble is that ripping the hydrogen out of water takes energy. Plants do it by means of photosynthesis, which makes use of sunlight as a source of energy. Scientists are searching avidly for some way of doing the work of photosynthesis in the laboratory and doing it even faster and better. We could then turn water (plus sunlight) into hydrogen, burn the hydrogen, and get water again. The fuel would never be used up; it would last as long as the Sun does.

Our Cousin the Coelacanth

Water is a milder, better arena for life than land is, or ever was. The proof is that so many forms of land vertebrates go back to a watery milieu and become adapted to life there. We need only think of sea snakes, sea turtles, penguins, seals, manatees, dolphins, and whales.

On the other hand, water vertebrates seem to have moved out onto land exactly once, and that was 370 million years ago. It has never happened again. Land is just too undesirable.

About 370 million years ago, the dominant vertebrates in water were the fish. There were two main kinds of fish, the ray-finned and the lobe-finned. The ray-finned fish had fins of thin, fine skin stiffened by horny rays, which were excellent for moving through water.

The lobe-finned fish (which lived in fresh water) had fins consisting of fleshy lobes, just barely fringed with a bit of fin. These were not as good for moving about in water, but they had an advantage. When, during a drought, a pool became too small for comfort, a lobe-finned fish could move out onto land, stumping along on its lobes, perhaps finding a larger pool.

Slowly, the lobe-fins evolved so they were able to move about on land more easily and stay there for longer intervals. Finally, they became adapted to indefinite stays on land, at least as adults. But they still had to go back to water to lay eggs.

Eventually, a particular species of amphibian developed an egg that could be laid on land, and so became the first reptiles. The reptiles continued to develop and evolve into numerous forms. From some of these arose the birds, and from others the mammals (including, eventually, us). A particular line of lobe-finned fish, then, are the ancestors of all land vertebrates, including us.

In water, however, the lobe-fins were not very successful and could not compete with the ray-fins, which were far better swimming machines. About 150 million years ago, when the dinosaurs were still on the rise, all the lobe-finned fish died out. All that was left of them were their progeny, the land vertebrates.

Or so it was thought. On December 25, 1938, a trawler fishing off South Africa brought up an odd fish about five feet long. It was taken to a local museum, where a Miss Latimer called in a South African zoologist, J. L. B. Smith. He examined it and realized it was an astonishing Christmas present. It was a lobe-finned fish and it had been alive when it was caught, though it was dead by the time it was pulled to the surface.

A branch of the lobe-fins had managed to adapt to salt water and had moved out to moderate sea depths. They had escaped extinction but were few in number and remained well out of the way, so that zoologists had never been aware of their existence. These lobe-fins of the salt sea were called "coelacanths."

The coelacanths were not the same line of lobe-fins that had emerged on land and developed into amphibians, reptiles, and all the rest. They were a related branch that had stayed in the water. The coelacanths are, therefore, not descendants of our ancestors but a collateral line. If not our grandfathers, they are our cousins, and of all the fish of the sea, they are the ones to which we are most closely related.

Other coelacanths have been caught. In fact, about ten to fifteen a year are hauled from the depths. Because they are not adapted to life near the surface, they are usually dead or dying by the time they are brought into view. As a result, we have not yet had the chance to study them alive, and certainly not in their own surroundings.

A German zoologist, Hans Fricke, now says that he has managed to observe live coelacanths by using a small submarine to dive near the Comoro Islands in the Indian Ocean, where the coelacanths live.

Fricke is still busy studying the photographs he took and hasn't yet made them public. He says, though, that the coelacanths move slowly (they're still poor swimmers, apparently). Whereas ray-fins use their tails for propulsion and their fins chiefly for balancing and turning, the coelacanths use their fins as paddles. They "row" through the water.

This is not as efficient as the ray-fin method, but it means that on land the coelacanth, and, presumably, other lobe-finned fish, would find walking more natural, since their rowing is essentially the same motion as walking. They would thus adapt a lot more easily to life on land.

Fricke also says the coelacanths are so few that the catches now made may drive them to extinction. After they have survived this long, this would be a shame. Besides, we owe them a little more than that. After all, they are our cousins.

The Relentless Population Rise

About the beginning of 1987, perhaps a little before, perhaps a little after, the population of the Earth reached 5 billion. The uncertainty arises from the fact that no accurate census is taken in many regions of our planet, so population estimates must be used.

The number, 5 billion, is large but it may not strike you as catastrophic. However, think of this. Modern man first appeared on Earth about 50,000 years ago, and the human population on Earth did not reach 1 billion until 1810. The 2 billion mark was reached around 1925. In other words, it took 50,000 years to build the human population to 1 billion, another 115 years to add a second billion, and only 30 more years to add a third billion by 1955. Then, 21 more years was enough to add a fourth billion, and no more than 10 more years was required to add a fifth billion. It would probably take only 9 more years to add a sixth billion.

It is quite obvious that population is going up faster and faster. This is not surprising since the more people there are, the more babies there are. In addition, as civilization has developed, the lives of people have gradually grown more secure and the death rate has dropped. This has been particularly true in the last century and a half, thanks to modern science and medicine. (What counts is not how many babies are born, but by how much the number of births exceeds the number of deaths.)

This is dangerous. The relentless rise of population means there must be more and more room for human beings, so that the wilderness is being wiped out. To support the growing population, more and more resources must be wrenched from the Earth, and more and more pollution of one sort or another must be produced. We are destroying our planet.

How long can this continue? Not very long. The Earth is not likely to support many more people. I have heard raving optimists say that with the advance of science and with an end to the waste of war and war preparations, Earth might well be able to support a population of 50 billion. I doubt this very much, but even if it were true, at the present rate of increase, Earth will reach a population of 50 billion in just over a century: in 2100. Then what?

Might we not ship the surplus population to the Moon or Mars or let them live on artificial worlds in orbit about the Earth? If so, this means that in the next century, we would have to relocate 45 billion people in space merely to keep Earth's population at its present 5 billion. Does anyone really think we can do that?

In fact, let's carry matters to an absolute extreme. All the matter in the universe is thought to total something like 200 billion billion billion billion tons. Suppose we imagine that we can somehow overcome all obstacles, send human beings like a flash into every corner of the universe, and turn all the stars and planets into nutritious food and breathable oxygen. We can then eat everything and the population can increase to 4,000 billion billion billion billion people. How long would that take? A billion years? A trillion years?

Well, it won't. At our present rate of increase, if we can imagine its continuing indefinitely, it would take only thirty-five hundred years.

We shall have converted the whole universe into human beings by A.D. 6500. Obviously, we're not going to do that, so the increase will have to stop—and soon! But how?

Actually, there are only two ways in which we can stop a

relentless rise in population. First, we can raise the death rate until the number of people dying is greater than the number being born. Or, second, we can lower the birthrate until the number of people being born is smaller than the number of people dying.

Nature's way of curtailing population is to raise the death rate. As the number of any species rises, there comes a point where famine takes over—and disease—and predators. In short, if we do nothing, the death rate will rise without our help. Humanity will suffer famine, epidemic, and war, and the population will be catastrophically reduced. In the process, civilization may well be wiped out. It is doubtful whether any sane person would consider this a suitable solution to the problem.

The alternative, as I've said, is to reduce the birthrate. Voluntary abstention from sex would do the trick, but it is not reasonable to expect such a thing to happen on a large scale. Failing that, people must learn to use contraceptive devices of one sort or another. Such use is being encouraged to a greater and greater extent in various parts of the world. In China, for instance, the birthrate has dropped drastically. The birthrate the world over has receded from its 1970 peak, which is a promising sign, but not by enough.

There are those who consider contraception immoral, but in that case, are mass death and destruction of civilization moral?

II

FRONTIERS OF SCIENCE

The Brightest X Rays

It is now possible for scientists to take pictures of molecules in one-ten-billionth of a second. This is so short an interval that the atoms making up the molecule have no time to move very far and are caught, so to speak, in midmotion.

The origins of the technique date back about three-quarters of a century, when it was realized that X rays were made up of very tiny waves. The waves are so small they can slip between the atoms in crystalline substances.

The atoms in crystals are arranged in an orderly fashion into various rows, tiers, and layers. In 1912, a German scientist, Max von Laue, showed that if X rays were allowed to impinge on a crystal they would bounce off these orderly arrangements of atoms and be *diffracted*, that is, curved away from their original path. If, after passing through the crystal, they struck a photographic plate, they would produce a symmetrical pattern of dots, depending on how they were diffracted by the different layers of atoms. From this *X-ray diffraction pattern*, scientists could calculate the positions of the atom layers and the structure of the crystal.

This could be done, eventually, not only for crystals of simple substances such as salt but also for orderly patterns of complex molecules such as proteins and nucleic acids. In this way, scientists learned about the atomic structure of hemoglobin (the oxygen-bearing matter that gives blood its color) and of deoxyribonucleic acid (DNA, the carrier of inherited characteristics).

There was a catch, though. The ordinary way of generating a beam of X rays was to have a stream of speeding electrons hit a metallic object. The sudden slowing of the electrons on collision produced the X rays. However, such X rays were weak and so dim that, in order to get good diffraction patterns, one often had to expose an object to the X rays for hours or even days.

During all that time, the atoms within the objects are moving about rapidly, vibrating about some point. The diffraction pattern therefore tended to show only that point, and there was no way of telling what the atoms were doing in the course of their vibrations. This was a particular loss in the case of complex molecules. Also, such a long exposure increased the chance that an X ray might, here and there, damage an atom and alter its arrangement, especially in connection with proteins and other molecules of very complex and fragile atomic structure.

There is another way of obtaining X rays. Electrons are made to whirl in circles under the influence of a strong magnetic field. It takes a great deal of energy to force the electrons out of their straight-line path, and this energy spills outward in the form of X rays—a very powerful beam, too.

In recent years such whirling electron beams have been used to produce ultrabrief flashes of ultrastrong X-ray beams. In the autumn of 1987, scientists at MIT were able to use a flash of X rays lasting only a few thousandths of a second to form a diffraction pattern of hemoglobin. This was better than had ever been managed before, but it still wasn't fast enough to "freeze" the molecule into a motionless state.

Now, however, beams of X rays have been produced that are so powerful that a complex molecule of importance to living tissue need be exposed to them for only a ten-billionth of a second. So powerful a beam is produced by making the electrons do more than turn in a simple circle: they wiggle back and forth as they do so. This modified device, called an *undulator*, has been successfully used at Cornell University.

Such powerful beams will pinpoint the position of each atom in midvibration with considerable precision. Nor will there be time for X rays to do any damage to the atoms. If diffraction patterns are then taken at different times, the atoms will be seen in somewhat different positions, and the various wrigglings of the molecule can be determined. This may give us new and unprecedented insights into how such molecules work within living cells.

Of course, the catch (there is always a catch) is that an undulator takes up a lot of room and is very expensive. This kind of ultrafast X-ray work cannot be performed in every laboratory; it will be confined to a few high-tech centers.

Scientists are now planning to build, perhaps by 1995, a device made up of as many as thirty-five undulators, each one more powerful than the one that is now being used at Cornell.

The ultrapowerful X rays thus produced ought to be ideal for studying, for example, the structure of the new superconducting materials and helping scientists determine just what atomic arrangements are needed to obtain superconductivity at still higher temperatures. They could be used to study the structure of meteorites to get more precise ideas of the original chemical makeup of the solar system. They could be used to determine the presence of various tiny impurities in materials where such impurities are not wanted—or where such impurities are necessary. And much more.

Reward Delayed

Half of the 1986 Nobel Prize for physics went to a German electrical engineer named Ernest A. F. Ruska. What did he do to deserve the prize? He built the first working electron microscope.

Ordinary microscopes, the kind that you and I have looked through in our time, make use of light waves. These light waves are focused by the lenses of the microscope in such a way that very tiny objects, too small to be visible to the unaided eyes, are enlarged so they can be examined in detail. However, they can only be enlarged so far. A light microscope can only make visible those objects that are at least as large as a light wave. A light wave simply skips over an object smaller than itself and cannot make it visible.

Suppose you focused X rays instead. X rays have tiny waves a thousandth or less the size of light waves. You could then see objects a thousand times or so smaller than ordinary microscopes make visible. The trouble is that X rays are too *energetic*. They smash right through the tiny objects they should make visible and you don't see them.

In 1924, however, it was shown that electrons (which everyone then thought of as tiny particles of matter) actually showed the properties of waves as well. Using electronic fields in the proper manner, electrons could be focused just as light waves could be. The electron waves, moreover, were no bigger than X rays but wouldn't push through matter in the same hardheaded way that X rays would. Electrons bounced off matter as light does, and we could "see," by tiny electron waves, much more than by the much longer light waves.

Seven years later, in 1931, Ruska built the first microscope that could actually "see" with electrons. But in that case, why didn't Ruska get the Nobel Prize then? Why did he get it in 1986, fifty-five years later, when he was eighty years old (and, fortunately, still alive)? To be sure, the first device was crude and didn't work as well as ordinary microscopes, but it established the principle and would surely lead to enormous advances. Why the delay, then?

When Alfred B. Nobel died in 1896 and left the money for an annual award of prizes, he intended to have scientists honored each year for something important they had done that year. It turned out, however, that it wasn't always easy to determine exactly what discovery in a given year would turn out to be important.

Sometimes, something looked great until it slowly fizzled. In 1908, for instance, a French physicist, Gabriel J. Lippmann, received a Nobel Prize for a system of color photography that eventually came to nothing. In 1903, a Danish physician, Niels R. Finsen, received a Nobel Prize for a way of treating skin disease with light that turned out to be totally unimportant. In 1926, another Danish physician, Johannes A. G. Fibiger, received a Nobel Prize because he found that certain cancers could be caused by a parasitic worm, but it later turned out the worm had no relation to the disease.

Those who awarded the Nobel Prizes learned by experience that it was better not to jump the gun. No matter how exciting a discovery might seem, it was better to wait and make sure it fulfilled its promise. Sometimes, of course, a discovery seems surefire and the judges can't wait. In 1956, two young Chinese physicists discovered that something called "parity" need not be conserved. This upset certain very basic rules of physics, and they got their Nobel Prizes the very next year—and they proved well deserved.

In general, there's a wait, though. Albert Einstein, in

1905, explained the photoelectric effect by quantum theory, thus helping enormously to establish one of the two great physical insights of the twentieth century. He got a Nobel Prize sixteen years later, in 1921. (By that time he had established the validity of his other great physical insight, relativity, an even greater feat, but that was not why he got the prize.)

In 1911, the American physician Francis P. Rous discovered a virus that seemed to produce cancer. It was a useful and important discovery, but it was not till 1966 that he won the prize for it, another fifty-five-year wait, like Ruska's. Rous was eighty-six years old at the time.

Sometimes, of course, scientists are *not* long-lived and never receive a Nobel Prize, although their work eminently deserves it. In 1914, an English physicist, Henry G. J. Moseley, worked out what we call "atomic numbers." This made sense out of the chemical elements and was absolutely sure material for a Nobel Prize. In fact, several scientists, who followed up Moseley's work in subsequent years, received Nobel Prizes for what they did.

Moseley himself, however, did not get his prize, for a simple if tragic reason. In the same year as Moseley's discovery, World War I broke out and Moseley at once enlisted. In 1915 he was killed at Gallipoli. He was twenty-seven years old, and one of the finest young minds of his time was snuffed out uselessly in a mismanaged battle: a symbol of the folly and futility of war.

The Noblest Element

Atom combinations make up everything we see about us on Earth, but some atoms are more reluctant to enter into combinations than others. In early 1988, however, an American chemist named W. Koch showed that even the least sociable atom may be coaxed into combination.

The atoms that are least likely to associate are the group of elements known as the "noble gases" (they are called "noble" because the attributes of standoffishness and exclusiveness are associated with nobility).

There are six noble gases, which, in order of increasing size of atom, are helium, neon, argon, krypton, xenon, and radon. None of these combines with other atoms under ordinary conditions. They exist as single atoms only.

In fact, the atoms are so indifferent to the presence of other atoms of their own kind that they don't tend to cling together even to the extent of forming liquids, so that none of them liquefies at ordinary temperatures. They are all gases and are found in the atmosphere.

The first noble gas to be discovered was argon, which was detected in 1894. It also is the most common, making up 1 percent of the atmosphere. The others were discovered a few years later; they exist on Earth only in small amounts.

Atoms combine with each other when one atom donates or shares electrons with another. The noble gases do not do this because their electrons are so symmetrically positioned within their atoms that any change requires a large energy input that is not likely to occur.

A large noble gas atom, such as radon, has its outermost electrons (the ones involved in chemical combination) far away from the nucleus. Therefore, the attraction between the outermost electrons and the nucleus is comparatively weak. For that reason, radon is the least noble of the noble gases, and the one that is most likely to be forced into combination by chemists who set up the right conditions for it.

The smaller the noble gas atom is, the nearer the outermost electrons are to the nucleus; those electrons are held in place more strongly, making it harder for the atom to form a combination with another.

As a matter of fact, chemists have forced the large-atom noble gases—krypton, xenon, and radon—into combination with such atoms as fluorine and oxygen, which are particularly eager to accept electrons.

The smaller-atom noble gases—helium, neon, and argon—are small enough and therefore noble enough not to be forced into combination by anything chemists can so far do.

The noble gas with the smallest atom is helium. Of all the different types of elements, it is the least likely to form combinations. It is the noblest element of them all. It is so reluctant to associate even with other helium atoms that it doesn't turn into a liquid until a temperature of only 4 degrees above absolute zero is reached. Liquid helium is the coldest liquid that can possibly exist and is crucially important to scientists for the study of such low temperatures.

Helium is present in the atmosphere only in tiny traces, but when radioactive elements such as uranium and thorium break down, they form helium. This accumulates in the ground, and certain oil wells yield helium as well. This is a limited resource but it has not run out yet.

Each helium atom has only two electrons, which are held so tightly by the helium nucleus that ripping away one of those electrons takes more energy than is required to remove an electron from any other element. With such a tight hold, can a helium atom be made to give up an electron, or share it, and form a combination with any other atoms?

To calculate the behavior of electrons, chemists use a mathematical system called "quantum mechanics," which was formulated in the 1920s. Koch, the chemist, applied its principles to the matter of helium.

Suppose, for instance, that a beryllium atom (with four electrons) combines with an oxygen atom (with eight electrons). In the combination, the beryllium atom hands over two electrons to the oxygen atoms and they cling together as a result. Quantum mechanics shows that the side of the beryllium atom facing away from the oxygen is very electron-poor as a result.

If a helium atom comes along, it will, according to the quantum-mechanical equations, share its two electrons with the electron-poor end of the beryllium atom. The combination helium-beryllium-oxygen will be formed.

So far, no other atom combinations seem to yield the right conditions for trapping helium, and even helium-beryllium-oxygen will probably cling together only at temperatures cold enough to liquefy air. It is now necessary for chemists to work with materials at very low temperatures to see whether they can actually make practice confirm theory, trapping helium into combination and thus defeating the noblest atom of them all.

Raising the Temperature

Scientists don't expect to find perfection, but sometimes they do. In 1911 a Dutch physicist, Heike K. Onnes, was lowering the temperature of mercury toward absolute zero. *Absolute zero* is the lowest possible temperature and it is equal to -273 C or -459 F. Onnes was testing the manner in which mercury conducted electricity at very low temperatures. He expected its resistance to electrical flow to decrease in an orderly way as the temperature lowered.

That didn't happen. At 4.12 degrees above absolute zero the resistance suddenly disappeared altogether. Mercury's electrical conductivity became *perfect.* Any electric current set up in a ring of frozen mercury at a temperature of less than 4.12 degrees above absolute zero would simply keep on flowing, undiminished, forever. This was called *superconductivity.*

Superconductivity was found in other elements when they were cooled all the way down. Some only became superconductive at still lower temperatures than mercury did, some at rather higher temperatures. The record high for an element is that of the radioactive metal technetium, which becomes superconductive at 11.2 degrees above absolute zero.

Superconductivity has more than theoretical interest. If electricity could be carried along cables under superconductive conditions, there would be no loss to resistance, and that would save billions of dollars. It is also possible to use superconductivity to produce very strong magnets, which would make it crucially important in setting up huge atom-smashing machinery. Superconductivity would also be useful in advanced computers and in many other aspects of current high technology.

There is one catch, though. In order to maintain a solid at so low a temperature, it must be kept immersed in a liquid that boils at that temperature. The liquid can't warm up past its boiling-point temperature; it simply boils off slowly. If more of the liquid is added, the very cold temperature can be maintained without difficulty.

As it happens, at temperatures below 14 degrees above absolute zero, there is only one liquid that can exist: liquid helium. Everything else, even the air about us, is frozen solid at such temperatures.

Liquid helium boils at 4 degrees above absolute zero. Anything immersed in slowly boiling liquid helium remains at 4 degrees above absolute zero indefinitely. However, helium is a rare substance, and it is very difficult to keep liquid helium cold enough to prevent it from boiling off rapidly. This seriously limits the uses of superconductivity.

The next coldest liquids are liquid hydrogen and liquid neon. Hydrogen is liquid between 14 and 20 degrees above absolute zero. Anything immersed in slowly boiling liquid hydrogen will remain at 20 degrees above absolute zero indefinitely. Neon is liquid between 25 and 27 degrees above absolute zero. Anything immersed in slowly boiling liquid neon will remain at 27 degrees above absolute zero indefinitely.

Hydrogen is much more common than helium, but hydrogen vapors are explosive. Neon is relatively rare, but it is more common than helium, and, like helium, it and its vapors are completely inert and cause no trouble. Maintaining either hydrogen or neon in the liquid state is much easier, and far less expensive, than maintaining helium in the liquid state.

For a long time, therefore, it has been the ambition of physicists to find some substance that is superconductive at liquid hydrogen temperatures. Pure elements simply won't do, but there is an alternative.

When solid elements (usually metallic) are mixed, frequently the mixture, or alloy, has properties that are not quite like those of any of its separate components. Scientists, when studying alloys, found some that were superconductive at temperatures greater than those for any pure element. In 1968, an alloy of niobium, aluminum, and germanium was found to remain superconductive at 21 degrees above absolute zero. Over the next eighteen years adjustments in the percentages of the mixture were studied, and in 1984, a niobium-germanium alloy that was superconductive at 24 degrees above absolute zero was found. Liquid-hydrogen superconductivity had become possible, but just barely.

And then in the last days of 1986, there were not one but two surprise announcements. At the University of Houston, it was reported that an alloy of lanthanum, barium, copper, and oxygen was superconductive at 40 degrees above absolute zero—but there was a catch. In order to maintain the superconductivity at so high a temperature, the alloy had to be kept under a pressure of hundreds of thousands of pounds per square inch. At Bell Laboratories, however, an alloy was reported that was fully superconductive at 36 degrees above absolute zero and remained so under ordinary conditions. No pressure was required. This makes it look as if liquid hydrogen superconductivity is on its way toward practicality.

Possibly, still higher superconductive temperatures will be achieved. Temperatures of 78 degrees above absolute zero may be theoretically possible, which would bring matters to the point that liquid nitrogen could be used (nitrogen being both common and safe). The ideal is superconductivity at ordinary temperatures, and even that may be within reach someday.

Atom Smasher Supreme

They were called atom smashers in the beginning. Later, they were given the more scientifically accurate name of "particle accelerators," but drama is drama. To the public they are still atom smashers.

The idea of particle accelerators is just what the name implies. Tiny subatomic particles, carrying an electric charge, are made to move faster and faster by magnets. Eventually, when they have been accelerated to the highest speed possible, the particles are smashed into a target.

The faster the particles go, and the more massive they are, the harder their collision and the greater the energy produced. The energy is partly converted into mass, and thus produces new particles, some so massive that they would not ordinarily be observed in nature.

The first particle accelerator was constructed in 1928 and produced speeding particles with energies of nearly 400,000 electron-volts. Such instruments had the particles accelerating in a straight line so that to reach greater and greater energies, one had to make the instruments miles long and they quickly became unwieldy.

In 1931, Ernest O. Lawrence of the University of California had the brilliant idea of making particles move in a spiral between the poles of a magnet. He called the device a "cyclotron." In such instruments, particles could move a long distance in spirals without requiring much room. In fact, his first cyclotron was only a foot across and cost very little money, but it could produce particles with an energy of 1.25 million electron-volts.

At once the move was on to build larger cyclotrons and get more energy out of them. By 1939, the University of California had one five feet across that was capable of producing particles with energies of 20 million electron-volts.

The design of such devices was continually improved, and by the end of World War II there were particle accelerators that produced energies of 200 to 400 million electron-volts. By 1949, energies of up to 24 billion electron-volts were generated.

Such energetic devices could produce quantities of "antimatter" particles, for instance, antiprotons. These had been predicted in theory but had never been observed until enough energy was produced to *form* them.

But in order to make really powerful atom smashers, the particles being accelerated had to be swung about in larger and larger curves, making the devices of today miles across. They require enormous and powerful electromagnets. To be powerful enough the electromagnets had to be kept superconductive so that the electrical current producing the magnets suffered zero losses. This can be done only if the conductors are kept in liquid helium within 4 degrees of absolute zero.

Particle accelerators, with all this taken into account, have now become enormously expensive and nobody can find enough money to build them except governments, sometimes combinations of governments.

The largest accelerator in the United States is in Batavia, Illinois, and it is four miles in circumference. But this is by no means the largest in the world. The Soviet

Union and a group of Western European nations each have larger ones and are designing new ones that are still larger. The Soviets are planning one thirteen miles around and a Western European consortium is planning one with a sixteen-mile circumference.

The recent discoveries in subatomic physics have come from Europe, and the United States is fearful of losing its edge in basic science. There still remain important particles to produce. There is a sixth quark, predicted but not yet found. There is something called a Higgs particle, which is theoretically important and which has eluded us. There are other unfound particles such as magnetic monopoles, and perhaps still others totally unexpected.

The United States is therefore planning to build a particle accelerator that will have a circumference of fifty-two miles and will cost no less than $6 billion. It will ultimately produce particles with energies twenty times as great as those generated by any other accelerator in existence. These particles will be sent in opposite directions and smashed together head-on.

Is there any point to finding rarer and rarer particles at higher and higher energies? Yes, for when the universe began with the big bang, incredible energies were involved, energies we can never possibly reach. The higher energies we produce, the more likely we are to be able to deduce what happened at the start and the more we will know about the universe.

We are not likely to be able to go much higher, however. Spending more than $6 billion for still larger machines may be difficult to manage. Some scientists worry that spending this kind of money on one device is going to starve other areas of science that are also crucially important.

The Two Nothings

The 1988 Nobel Prize in physics was given to three Americans, Leon Lederman, Melvin Schwartz, and Jack Steinberger, for work they did with subatomic particles that come as near to nothing as it is possible for anything to come.

The particles are called "neutrinos." They have no mass. They have no electric charge. They are so indifferent to matter that they can pass through a trillion miles of solid lead and only a very few of them will be stopped. It was these nothing particles that our prizewinning physicists worked with, back in the early 1960s.

They were interested in the *weak interaction,* one of the four ways that particles can interact with each other. Gravitational interaction holds the universe together; the strong interaction holds atomic nuclei together; the electromagnetic interaction holds atoms and molecules together; and the weak interaction allows some nuclei to break down.

It was very difficult to study the weak interaction, and it occurred to Melvin Schwartz that the way to do it was to make use of beams of neutrinos. Neutrinos are involved only with the weak interaction. This is the chief reason they can pass through ordinary matter as if it isn't there.

But how does one make a beam of neutrinos? One way is to begin with a beam of protons, which have mass and electric charge and can be accelerated to high energies easily. Such a proton beam can be made to smash into matter and produce a thick spray of energetic particles. Among these particles are "pions," which quickly break down to another type of particle called "muons" and to neutrinos.

Lederman, Schwartz, and Steinberger got together to work on such beams. They allowed the spray of particles to fall on steel armor taken from a battleship that was being dismantled. The armor was piled up until it was thirty-three feet thick. It stopped all the particles but the neutrinos. On the other side of the armor, the experimenters had a beam made up only of neutrinos.

For the researchers to use this neutrino beam and to figure out the details of how the weak interaction worked, the neutrinos had to be absorbed by matter and produce changes. Neutrinos go right through matter, but not always. In a beam of trillions of neutrinos, most may pass through matter, but a few dozen will be stopped.

Because of this, it is possible to study the neutrinos themselves. The first thing we need to know is how various types of neutrinos differ from each other, if at all.

Neutrinos form in two ways. Whenever a muon is formed from a pion, a neutrino is also formed. Whenever a muon breaks down to an electron, another neutrino is formed. Thus, there are two neutrinos: a "muon neutrino" that accompanies muon formation and an "electron neutrino" that accompanies electron formation.

The muon itself is identical to an electron in every known way but one. The muon is about two hundred times as massive as an electron; it is a "heavy electron." The muon neutrino and the electron neutrino, however, do not have even that much difference. Both were precisely the same in every measurement that physicists could make. Did that mean they were actually identical particles?

The three physicists decided to try to settle the matter by what is now called "the two-neutrino experiment," using the neutrino beams they had learned to produce. The neutrino beams were made up of muon neutrinos, for they had all been formed along with muons. Now, when these neutrinos were absorbed by matter, they ought to form muons.

If muon neutrinos were separate particles, distinct from electron neutrinos, then only muons should be produced. On the other hand, if muon neutrinos and electron neutrinos were the same particle, then the neutrino beam ought to produce both electrons and muons, and, in all likelihood, in equal quantities.

For eight months, Lederman, Schwartz, and Steinberger kept bombarding matter with neutrino beams. Countless hundreds of billions of muon neutrinos smashed into the matter, and in those eight months, just fifty neutrinos were stopped. Every one of them produced a muon.

This made it clear that muon neutrinos and electron neutrinos were distinct and different particles, but even today physicists don't know just what it is that *makes* them different. All of their properties that can be measured seem identical, but even though scientists can't tell them apart, other subatomic particles can, somehow. This means there are two nothings, two different nothings.

In fact, things are worse than that, for about a dozen years after the two-neutrino experiment, a third neutrino had to be taken into account, a "tauon neutrino." This, presumably, is different from each of the other two, leaving us faced with three nothings that differ from each other in some way we can't detect.

Nevertheless, this undetectable difference is important in preparing subtle theories of the basic structure of matter. The two-neutrino experiment is well worth a Nobel Prize.

Supercritical

QUESTION: What is neither liquid nor gas, but a little bit of both?

ANSWER: "Supercritical fluid," and scientists are now learning to make good use of it.

Ordinarily liquids and gases are quite different. A liquid has a definite volume; you can half-fill a container with liquid. A gas does not have a definite volume; it always fills a container completely.

A liquid can dissolve solids and other liquids, but a gas can't.

A liquid is much more dense than a gas. Liquid water is 1,250 times as dense as gaseous water (steam). A quart of water, in other words, weighs 1,250 times as much as a quart of steam.

You can change a liquid into a gas with heat. Thus, if you heat water, you eventually bring it to its boiling point and it bubbles away as steam. This boiling point, under ordinary conditions at sea level, is 100 Celsius (C) or 212 Fahrenheit (F).

If, however, you want to prevent water from boiling at 100 C, you must put pressure on it in order to keep its molecules put, so to speak. As the temperature continues to go up, you must put water under higher and higher pressure to prevent it from boiling. Eventually, if the temperature is high enough, no amount of pressure will prevent it from boiling.

The temperature above which a liquid will boil, regardless of pressure, is the *critical temperature*. The critical temperature for water is 374.2 C (705.6 F). The *critical pressure*, which will just keep water liquid at that temperature, is 218.3 times ordinary atmospheric pressure.

Above that temperature and pressure, you have *supercritical water.* Like steam, it has no definite volume and will fill any container. It is, however, much denser than steam and is, in fact, one-third as dense as liquid water. Its most astonishing property, though, is that it will dissolve substances as liquid water does.

Every liquid has its own critical temperature and pressure, some higher than that of water, and some lower. This was first discovered in 1869 by an Irish chemist named Thomas Andrews. Carbon dioxide, for instance, has a critical temperature of 31 C (88 F) and a critical pressure of 72.85 atmospheres. Hydrogen has a critical temperature of −204 C (−400 F) and a critical pressure of 12.8 atmospheres.

Naturally, we don't find supercritical fluids in nature under ordinary circumstances on the Earth's surface, but they can exist at the center of planets where the temperatures and pressures are high enough. The interior of the giant planet Jupiter, for instance, is made up largely of supercritical hydrogen at a temperature of tens of thousands of degrees.

In the laboratory, scientists have learned to produce temperatures and pressures high enough to form supercritical fluids. At the University of Maine, chemical engineer Erdogan Kiran has designed a steel chamber within which pressures of up to one thousand atmospheres can be produced as well as temperatures that are high enough to produce supercritical fluids. You can even watch substances dissolve in the supercritical fluids through half-inch-thick windows made of a transparent, rocky synthetic substance.

Supercritical fluids, like ordinary fluids, will dissolve some substances more easily than others. Therefore, they can be used to extract some parts of a complex mixture and leave the rest behind. However, if the supercritical fluid is too hot, it can damage the molecules of the substance it dissolves, and even of the molecules it leaves behind.

Supercritical water is definitely too hot to be trusted to extract substances without harming them, especially substances that are "organic" and have large and rather rickety molecules. In that case, why not use supercritical carbon dioxide, which is much cooler and requires smaller pressures for its formation?

In West Germany, supercritical carbon dioxide has been used to extract caffeine from coffee beans. As it happens, supercritical carbon dioxide takes out only the caffeine, leaving everything else intact.

Ordinary liquid solvents tend to remove other ingredients along with the caffeine. What's more, traces of ordinary solvents that may be dangerous in the long run may remain.

But when the supercritical carbon dioxide is (with its load of caffeine) removed, none remains. After all, when the pressure is relieved, any remaining supercritical fluid just turns into gas and vanishes. Decaffeinated coffee made in this manner should taste just like the original.

There is hope that supercritical fluids can be used to perform other extractions efficiently and harmlessly. Perhaps oil can be extracted from potato chips, leaving a low-calorie product without impairing the taste. Or fish oils may have the source of their smell removed, leaving nutrition unaltered. Supercritical fluids also hold promise for the purification of medicinal products and for the study of protons, nucleic acids, and other complex molecules.

A Question of Priorities

Two hoped-for advances are on a collision course, and scientists are being presented with a dilemma with no easy solution, one that may risk billions of dollars or set back research for decades.

First, superconductivity. There have been exciting and unexpected new findings in the realm of superconductivity. Electric currents that had been able to flow without loss and without developing heat at extremely low liquid-helium temperatures suddenly appeared able to do so at considerably higher liquid-nitrogen temperatures, making the whole process cheaper and more practical.

Second, the "supercollider." Plans have already been made to build a new, huge particle accelerator, about seventeen miles across and fifty-two miles in circumference at a cost of billions of dollars. With it, scientists hope to learn new facts about the fundamental building blocks of matter and about the origin of the universe.

But here's the quandary.

In order to build the new particle accelerator, a great quantity of electricity to produce extremely powerful magnets stretching all around the fifty-two-mile circumference of the device must be used. These magnets serve to produce an electromagnetic field powerful enough to accelerate particles to nearly the speed of light, forcing some to smash into others in ways that will produce collisions of enormous energy.

In order to bring this about, the magnets must be

cooled to very low temperatures and made superconductive. In this way, without loss of current or development of heat, magnets that will be much more powerful than would be possible otherwise can be produced. This means the new supercollider must use a great quantity of costly liquid helium and a great deal of expensive machinery to maintain the helium in liquid form for as long as possible.

Some scientists have questioned the advisability of such a device on the grounds of expense. Not that the device isn't likely to produce important new knowledge obtainable in no other way, but there is only so much money that can be devoted to science. If the supercollider drains off billions of dollars, there may be little money available for other types of research. The loss of new knowledge in other fields may be greater in sum than the gain in subatomic physics. This is a hard case to argue because we don't know what the new knowledge gained in one case, or lost in another, may be.

However, the new developments in superconductivity offer those who object to the expensive new supercollider a powerful argument. They suggest that physicists wait because it will soon be possible to use new materials that will allow superconductivity at liquid-nitrogen temperatures. Liquid nitrogen is much cheaper than liquid helium and is much less difficult to keep liquid. In this way, the cost of the new machine will be cut 10 to 15 percent.

In fact, it may even be possible before long to have some materials that are superconductive at still higher temperatures and to construct magnets far more powerful than we can hope for now. More powerful magnets will create stronger fields that will bend the paths of speeding subatomic particles more sharply. Instead of being able to bend them only slightly, so that they have to speed about a circle seventeen miles in diameter in a path fifty-two miles long, they might be bent sharply about a circle less than two miles in diameter with a path only five miles long.

In that case, the land required for the machine will be reduced to only 1 percent of the area now needed, and the materials required will be similarly reduced. The money needed for the device will then be cut enormously, and some billions of dollars will be saved, to be allocated to other forms of research without hindering subatomic physics.

It sounds good, but there are subatomic physicists who object. So far the new superconductive materials have been produced only as laboratory specimens. How long will it be before they can be produced in quantity and with the properties required for the manufacture of powerful superconducting magnets? There can be all sorts of difficulties and engineering hang-ups that will have to be solved. Even small problems that arise unexpectedly might take years to solve.

In other words, physicists might have to wait ten or fifteen years for new materials, and then political or economic conditions may make the billions of dollars needed for a supercollider unavailable. Physicists hesitate to take that chance. They have the money and are reluctant to let go.

Well, then, to wait or not to wait? That is the question.

Suddenly, Thallium Plays a Role

Some chemical elements, such as gold and oxygen, are household names; some, such as neodymium and lutetium, aren't. But every once in a while some element that only chemists have heard of suddenly makes the news. This has happened to the chemical element thallium.

Thallium was discovered in 1861 by the British physicist William Crookes. He was studying the wavelengths of light given off by heated minerals when he found a beautiful green line at a wavelength not listed for any element that was then known. He tracked it down and isolated a hitherto unknown element that he named "thallium" from the Greek word *thallos*, meaning "green twig," in memory of the green line that had set him on the track.

However, there didn't seem to be much one could do with thallium. It rather resembles lead in its properties. It is a bit denser than lead and it melts at a slightly lower temperature. And it is poisonous. In fact, the first use found for thallium (in 1920, nearly sixty years after it had been discovered) was as a rat poison.

But now there is the matter of superconductivity. Some substances lose all electrical resistance at very low temperatures, and this property can be crucially important in various branches of science and technology. Thus people speak of superconductivity in connection with magnetically levitated trains, more powerful atom smashers, smaller and faster computers, and controlled nuclear fusion.

Until 1986, however, no substance was known that was superconductive at temperatures higher than 23 degrees above absolute zero. This is very cold indeed, considering that ordinary room temperature is about 300 degrees above absolute zero and the coldest Antarctic weather is 200 degrees above absolute zero.

But people had been testing only metals for the purpose. In 1986, it occurred to two scientists in Zurich, K. A. Mueller and J. G. Bednorz, to try certain ceramic substances, and, behold, they found superconductivity at temperatures of 36 degrees above absolute zero and promptly won Nobel Prizes in 1987.

The ceramics Mueller and Bednorz studied were based on copper oxide. It seems that superconductivity depends on electrons' making their way along connected sheets of copper and oxygen atoms. However, in order to obtain high-temperature readings, other types of atoms have to be present also. The other atoms have to be of such elements as barium, yttrium, and lanthanum. In particular, it seems that atoms of one or another of a group of elements called the "rare earth elements" have to be present.

There was no way of being sure how any of these ceramic mixtures would work. Chemists mixed different oxides (including the crucial copper oxides) in different proportions and baked them at different temperatures for different amounts of time to see what would happen. It was "cookbook" chemistry, and the mixtures weren't reliable. A particular mixture might be superconductive at pretty high temperatures one time and flop miserably when the next batch was made. It all depended on how the particles of ceramic melted together under heat.

The highest temperatures permitting superconductivity remained less than 100 degrees above absolute zero. (There were occasional reports of higher temperatures, but these were apparently mistaken.) To be sure, even a temperature near 100 degrees above absolute zero is marvelously high compared to what existed only a few years ago, but it is still less than Antarctic. Scientists want superconductivity at higher temperatures still.

It occurred to an American chemist, Alan Herman, to try thallium atoms in place of the rare earth elements. Thallium atoms are about the same size as the rare earth atoms and slip into the same places in the molecular structure.

In May 1987, for the first time, a ceramic without rare earth elements was found to be superconductive, and at a temperature of 80 degrees above absolute zero.

The original recipe called for oxides of copper, barium, and thallium, but in early 1988, Herman added a bit of calcium to the mixture and obtained a superconductive temperature of 105 degrees above absolute zero. The mixture containing thallium was the first one to penetrate the 100-degree mark; nothing without thallium had done as well.

It seemed that a great deal depended on how many layers of copper-oxygen atoms there were between the thallium layers on the boundary. The first thallium ceramic had had a single layer of copper-oxygen atoms between the thallium layers; the later one with the higher superconductive temperature had two.

Obviously, it was important to try a ceramic with three copper-oxygen layers between the thallium boundaries, and when this was done a superconductive temperature as high as 125 degrees above absolute zero was obtained. It is uncertain how many more layers can be piled on, and there are theoretical reasons for supposing the temperature may not go up indefinitely. Still, if as many as ten layers of copper-oxygen can be sandwiched in along with the thallium and other elements, superconductive temperatures as high as 200 degrees above absolute zero may be obtained. The Antarctic-temperature barrier will have been broken.

Just the same, thallium remains very poisonous and may be too dangerous for industrial use. So although scientists may have to search for another substance that will serve the same purpose, thallium will have played its role in advancing our technological ability.

Breaking the Bond

If you stick a pin into a balloon, how long does the balloon take to explode? Not long at all, but the time can be measured by high-speed photography. After all, the rubber takes *some* time to split apart under pressure.

But suppose you take a molecule only four-billionths of an inch across and do the equivalent of sticking a pin into it. How long will the molecule take to fall apart? Far less time than a balloon takes to explode—and now scientists have been able to measure that time.

A molecule is composed of a group of atoms. The atoms stick together because the tiny electrons in their outer regions overlap when they get close enough to each other. This overlapping produces a stable situation that tends to be retained. To retain it, the atoms must continue to remain in close proximity. The result is what is called a *chemical bond.*

Two atoms forming a chemical bond don't remain still. At any temperature above absolute zero, atoms tend to move about in random fashion. They can't move freely when they're held by a chemical bond, but they keep trying, so to speak. Two atoms held by a chemical bond may move away from each other, but the bond draws them back together. They move away again and are brought back again, over and over. As a result, they vibrate in place. Each is like a base runner forever taking a lead off first base and forever being brought back by a throw from the vigilant pitcher.

The bond acts like a tiny spring. The farther the atoms move from each other, the more firmly the bond acts to bring them back. However, if for any reason the atoms move away from each other by more than a critical amount, the bond is overstrained, as a spring would be, and breaks. The molecule falls apart and the atoms are free.

As temperature rises, atoms tend to move apart too far to be held by the chemical bond. If temperature rises high enough, molecules are sure to fall apart. They also tend to fall apart if energy is added in other forms. The question is how long they take to fall apart once sufficient energy is added.

A group of chemists at California Institute of Technology, headed by Ahmed Zewail, answered that question for the first time in 1987. They worked with iodine monocyanide, a molecule made up of three atoms—iodine, carbon, and nitrogen—attached side by side. If enough energy is added, the iodine will break away, leaving only the carbon and nitrogen (a "cyanide group") linked together.

The trick is to supply the energy in an extremely brief time, just long enough to damage the bond, no longer. How long after that does the iodine atom take to separate from the cyanide group?

The chemists added energy by means of a very short pulse of light. This pulse of light will knock an electron out of the bond holding the iodine atom to the cyanide group, so weakening the bond (like a pin's weakening the rubber of a balloon) that the iodine atom pulls free (like the balloon's exploding). The pulse of light is very brief indeed: 60 millionths of a billionth of a second. It strikes and is gone, and chemists can then wait for the damaged bond to break.

But how can they tell when the bond breaks?

It so happens that the isolated cyanide group absorbs light of a particular type and then gives off light of another type. This process is called *fluorescence* and it can be easily detected. The intact iodine monocyanide does not fluoresce, so the appearance of fluorescence means that the bond has broken and the cyanide group has been formed.

It is therefore necessary for the researcher to fire a brief pulse of laser light at the iodine monocyanide and then fire a second pulse immediately afterward to see whether fluorescence can be detected. The process is then repeated, with the second pulse fired at ever shorter intervals after the first. Finally, the second pulse comes so soon after the first that there is no fluorescence—there is no time for the bond to break apart.

In this way, the researchers found that the time needed for the bond to break after it had been damaged was 205 millionths of a billionth of a second. For the bond to break, the iodine atom had to move 120 millionths of an inch away from the cyanide group. Is there any way to visualize the ultrabrief interval of time a chemical bond takes to break? Well, we can try.

Light moves at a speed of 186,262 miles per second, the fastest speed that is permissible in this universe of ours. This is so fast that a ray of light can circumnavigate the Earth in one-seventh of a second or travel from the Earth to the Moon in one and a quarter seconds, or from the Earth to the much more distant Sun in eight minutes.

How far then will light travel in 205 millionths of a billionth of a second? The answer is 1 four-hundredth of an inch. In other words, the ultrafast beam of pulsed laser light strikes the molecule and manages to get only 1 four-hundredth of an inch past it before the bond breaks.

A Dream Comes True

In 1865, a chemist solved a problem in a dream, and now, a century and a quarter later, scientists have finally checked the matter in the most direct possible manner—and, behold, the dream solution turns out to have been correct.

It came about in this way. In the early 1860s, chemists were learning the manner in which atoms combined with each other to form molecules. The system they used explained the properties of molecules in terms of their atom connections. A few simple rules about the manner in which the various kinds of atoms hook up to others produced models that looked like Tinker Toys and that clarified an enormous number of chemical observations.

Leading in this effort was a German chemist named Friedrich A. Kekulé, who still was stumped by one important problem the Tinker Toy models didn't seem to solve.

This involved a compound named benzene. Kekulé knew that each benzene molecule was made up of six carbon atoms and six hydrogen atoms, but there seemed no way of fitting them together properly. No matter how they were put together, the result was the kind of molecule that ought to be very active: that ought to combine with other atoms and molecules easily. Unfortunately, this was not how benzene behaved in real life. It was a very stable compound that combined with other atoms and molecules only with considerable difficulty.

As long as this discrepancy existed, the whole Tinker

Toy system was suspect, and chemists do not relish having to look for a new kind of model.

Kekulé spent years on this problem. He arranged the carbon and hydrogen atoms in every conceivable way but failed to find a satisfying model. The solution arrived in an unexpected manner. He had taken a horse-drawn bus that carried him slowly along the streets of Ghent, Belgium, toward the university where he was teaching at the time. He was tired and, of course, was thinking of the benzene problem, which consumed all his thoughts.

He fell into a doze and, even while he was sleeping, the problem did not leave him. He dreamed about chains of carbon atoms, twisting this way and that as they attached themselves to hydrogen atoms. And, in the dream, a chain of atoms suddenly curved so that one end hooked to the other end and formed a tiny hexagon of carbon atoms, spinning endlessly.

He woke with a start and realized he had the solution. Everyone had been taking it for granted that six carbon atoms would form a straight line with hydrogen atoms attached here and there. But what if the six carbon atoms formed a ring?

Back in his laboratory, he considered a molecule of benzene as consisting of a ring of six carbon atoms, with one hydrogen atom attached to each. Such an arrangement was very symmetrical and thus should give the molecule considerable stability. He considered the ways in which other atoms could attach themselves to such a ring and found that the prediction exactly matched the way the molecule behaves in reality. There were, for instance, just three ways in which two chlorine atoms could substitute for two hydrogen atoms, both in the model and in reality.

The six-carbon ring has been accepted ever since.

The ring in itself didn't quite account for the stability, to be sure, but in the early twentieth century it was found that atoms consist of tiny nuclei surrounded by light electrons. It is the electrons that interact with each other to form bonds between atoms. In 1939, Linus Pauling showed that, in the case of molecules like benzene, the electrons' interaction produces a very stable situation.

But though every chemical property of benzene discovered since Kekulé's time supported the hypothesis that each benzene molecule is a ring of carbon atoms shaped like a tiny hexagon, the evidence was all indirect.

Finally in 1981, a device called a scanning tunneling microscope was invented at IBM. It consisted of an extremely fine tungsten needle that emits electrons in a vacuum. These electrons bounce off the surface of material. From the reflection of those electrons, a computer can calculate the appearance of the reflecting surface. The surface can be seen in such detail that the atoms themselves can be discerned.

It would be interesting to bounce electrons from a surface made of solid benzene, but something that conducted electricity was needed, and benzene doesn't have the property of conduction. What's more, even in solid form, benzene molecules move about so much that the picture was too fuzzy to show much.

The benzene was combined with carbon monoxide to keep it steady, and the whole bound to rhodium metal, which conducts electricity. Pictures were obtained at last in 1988. And what they showed was carbon rings in the shape of hexagons. Scientists could finally see Kekulé's dream. It was correct.

Getting Old

Scientific language doesn't match ordinary speech sometimes. If you come across the expression "free radical," you are likely to think of it as describing some extremist who is not in jail. In chemistry, however, it has an entirely different meaning.

In chemical terminology, a *molecule* is made up of more than one atom. Each atom in a molecule is attached to other atoms by a pair of electrons. Thus, a carbon atom may be attached to four different hydrogen atoms by four different electron pairs. Under some circumstances, a hydrogen atom may break loose, taking its electron with it. What's left of the original molecule is a carbon atom with only three hydrogen atoms. Where the fourth hydrogen atom should be is a single electron unattached to anything.

A molecular fragment containing that single electron is a *radical*. That single electron is very active, tending to tear strongly at other molecules in order to grasp an atom with which it can make up an electron pair again. This happens so rapidly that a radical, even if formed, does not last long and may just snatch up the atom that broke loose before it can quite get away. Nevertheless, if a radical can last long enough to wander about a bit and seize an atom from some other molecule, it is spoken of, during its brief existence, as a "free radical."

Free radicals can form within living cells. Energetic radiation, such as cosmic rays, X rays, or ultraviolet light from the Sun, can produce them. So can certain chemicals. These free radicals can last long enough to damage neighboring molecules. When these damaged molecules happen to be proteins, enzymes, or, worst of all, the deoxyribonucleic acid (DNA) molecules in the genes, the cell suffers. Some portions of the cell machinery may be knocked awry.

The body has ways of preventing or correcting damage by free radicals. Substances such as vitamin C and vitamin E can give up electrons easily, and, in doing so, they can satisfy the appetites of the free radicals and prevent them from tackling other molecules. The body also has corrective mechanisms that can repair molecules damaged by free radicals.

However, not all free-radical damage can be prevented or repaired. This means that as life goes on, damage to the cells does take place and it does accumulate. With the years, more and more cells are crippled and various necessary portions of the body machinery become more and more rickety and inefficient.

There are some scientists who think it is this accumulating damage that causes old age and makes it certain we all die in the end, even in the absence of infection or accident.

If this is so, then we might live longer if we could find some more powerful means than the body itself supplies for preventing free-radical damage. For instance, there are some plants, such as the creosote bush, that have an unusually long life span. The creosote bush contains a plentiful supply of the compound nordihydroguaiaretic acid (NDGA, for short). This can short-circuit free radicals by handing them an electron and perhaps does so more efficiently than vitamins C and E do.

A biochemist at the University of Louisville, John P. Richie, Jr., has recently tested this possibility by feeding NDGA to female mosquitoes. Such mosquitoes normally live for an average of 29 days, but with NDGA they lived for an average of 45 days. That's an increase of 50 percent. If it worked similarly on human beings, it might increase our average life span from 75 years to 113 years.

It's unlikely that anyone will try to feed human beings NDGA as an experiment, but Richie's observation seems to support the free-radical theory of aging. There may be other, less troublesome, ways of preventing free-radical formation or encouraging their removal so that human life can be extended considerably.

But the question that must be faced is whether we would want to do this, if it turns out we can.

An extended life span for human beings will further speed up the population increase and make it necessary to lower the birthrate even further than now seems advisable. This means that there will be fewer young people. Government, business, all the machinery that runs society will be conducted for longer and longer periods by older and older people and what young people there are will have to wait longer and longer for their chance to take over. Does this matter?

It probably does. It is not just that young people are young; they are new. Every youngster represents a new gene combination that can produce a brain that might just be able to tackle problems in a new and creative way. Society in the control of the long-lived oldsters with an ever-slower infusion of the young and new might tend to decay and become static. In fact, it may well be that the death of the individual is necessary for the health of the species. The advantage that you and I might gain by living longer could be paid for by the general decline of humanity.

The Elusive Quark

Since the time of the ancient Greeks, thinkers have been trying to answer the question, What are the basic objects making up the universe? Scientists today are still trying to answer this, but the last bit of information continues to elude them. It seems to slip away.

For instance, the universe is made up of a number of elements, simple substances that cannot be made simpler by ordinary chemical methods. Scientists have identified more than one hundred of them. Each element is made up of atoms, objects so tiny it takes about 250 million of them lined up side by side to stretch across an inch. We can say, therefore, that the universe is made up of more than one hundred different types of atoms.

But are the atoms really what the universe is made of? Or are the atoms themselves made up of still smaller and simpler objects?

At the beginning of the twentieth century, scientists discovered that atoms have a structure. The outer rings of the atom contain electrons, and at the very center is an atomic nucleus, so tiny that it takes about 100,000 of them lined up side by side to stretch across a single atom. Electrons are all alike, no matter what kind of atom they are found in. Atomic nuclei are different. Every kind of atom has its own kind of nucleus.

However, atomic nuclei are, in turn, made up of two kinds of particles, protons and neutrons, and all protons and neutrons are the same regardless of which atomic nuclei they are found in. By the early 1930s, it really seemed as if the matter of the universe, in all its apparently infinite variety, might be composed of three types of particles: electrons, protons, and neutrons.

But matters got more complicated. Actually, there are three varieties of electrons, each associated with a different type of neutrino, and all six of these have a mirror image. This makes twelve electron-type particles (called *leptons*) all together, and each one is a *fundamental particle* that, as far as we know, cannot be broken up into anything simpler. For scientists, twelve leptons is not too many to handle.

The neutron and proton are different. In the first place, neutrons and protons are a little more than eighteen hundred times as massive as electrons, so that they represent about 99.95 percent of the universe. Then, too, scientists began to identify all sorts of other particles that were more massive than electrons, literally hundreds.

These were too many and seemed to make the universe so complicated that once again scientists had to wonder whether these particles were composed of still smaller and simpler ones.

In the 1960s, scientists postulated the existence of new particles that are the building blocks of the massive particles. They called these new particles *quarks*. Neutrons, protons, and other particles even more massive, they theorized, are made up of three quarks each. Particles heavier than electrons but lighter than neutrons and protons (so-called mesons) are composed of two quarks each.

It turns out that there are twelve different quarks, just as there are twelve different leptons, and the quarks, too, seem to be fundamental particles. Now we can say that the entire universe in all its infinite variety is made up of twelve different leptons and twelve different quarks and all the particles commonly found in nature are either leptons, quarks, or quark-combinations (plus certain particles called "bosons" that allow other particles to interact with each other). This seems simple enough, but one serious problem remains.

Scientists can very easily detect free leptons and can study them. But they cannot detect free quarks. The electron has an electric charge of one unit; so has the proton. All electric charges are multiples of that unit. Scientists have calculated, however, that quarks must have fractional electric charges. Can we be sure of that if we have never studied them in their free state?

In fact, we can't even be *sure* they exist and are not just mathematical conveniences. For instance, we know that a dollar is the equivalent of ten dimes, but that doesn't mean that if we tear up a dollar bill we will find ten metal dimes in the shreds. The dimes in a dollar are mathematical.

It would be easy to identify a free quark by its fractional electric charge, which doesn't exist in any other kind of particle, but those fractional charges have never been found. Some were reported in the late 1970s, but when the experiments were repeated, it turned out the report was mistaken.

Free quarks must exist under very extreme conditions: at the center of neutron stars, for instance, or immediately after the big bang. But how can scientists reproduce such extreme conditions in the laboratory?

There is a little hope here. It may be possible to force particularly massive atomic nuclei into collisions at enormous energies, if we build powerful enough atom-smashing machines. The shattering nuclei may then liberate, just briefly, individual quarks. When that happens, scientists may possibly catch a glimpse of these elusive fundamental components of the universe. *Possibly.*

How Many Particles?

How many different particles make up the universe and everything in it? How many remain to be discovered? Physicists are now on the edge of getting some answers to those important questions.

There are three different classes of fundamental particles (that is, particles that can't be broken down into anything simpler). They are (1) *leptons*, (2) *quarks*, and (3) *bosons*.

The most important lepton is the "electron," which is found everywhere. There is a heavy electron called a "muon"; although it doesn't exist in nature in appreciable quantities, it can be made in the laboratory. There is a still heavier electron called a "tauon." Each of these particles has a neutrino associated with it, and all three neutrinos are different. That makes six leptons altogether.

There is also "antimatter," which is just like ordinary matter but has opposite characteristics such as electric charge. Antimatter doesn't exist in the universe in appreciable quantities, but it can be made in the laboratory. Antimatter is made up of six different "antileptons." That makes twelve leptons and antileptons.

The quarks come in six varieties also. The most important of these are the "up-quarks" and "down-quarks," which are the lightest ones. They make up the protons and neutrons that are found everywhere. The heavier a particle is, the more difficult it is to form. The most massive quark, "the top-quark," is 8,000 times as massive as the lightest quark and it hasn't been formed yet, but scientists are sure that it exists. For each quark there is an "antiquark," so there are twelve quarks and antiquarks altogether.

The bosons are particles that make it possible for leptons and quarks to interact with each other. There are four kinds of interactions: gravitational interaction, for which there is one boson; electromagnetic interaction, for which there is one boson; the weak interaction, for which there are three bosons; and the strong interaction, for which there are eight bosons. That means thirteen bosons altogether.

Leptons, antileptons, quarks, antiquarks, bosons equal thirty-seven particles altogether.

Are these all there are? Well—

The most massive of the weak interaction bosons is one called the Z degree particle. It is twice as massive as the most massive quark and it wasn't until 1984 that it was first formed and observed. The man in charge of the work that found it was the Italian physicist Carlo Rubbia, for which he was awarded a Nobel Prize.

A very massive particle is formed by forcing two ordinary particles together with great force. The particles smash each other into a spray of other particles. The energy of collision can be converted into mass so that the particles formed can be much more massive than the particles that originally collided.

Right now, Carlo Rubbia is working near Geneva, Switzerland, with the "Large Electron Positron" collider, usually referred to as the LEP. In it, a stream of electrons is whirled in a circle in one direction and a stream of positrons (the antiparticle equivalent of electrons) is whirled in the same circle in the other direction. They smash into each other head-on and, in that way, form other particles. If the energy of collision is just right, they form Z degree particles.

Construction of the LEP was begun in 1981. Electrons

and positrons were to be made to pass through a circular tube nearly 17 miles in circumference. The tube was to contain a vacuum so that there would be no air molecules to collide with. The LEP finally went into action in July 1989 and within four weeks had formed its first Z degree particle.

The LEP is not the first device to form Z degree particles. In the United States two different devices have been used to form them. However, the LEP has the capacity to fine-tune the quantity of energy it produces so that it is just right for Z degree particle formation. This means that when it really gets going it should produce them in quantity. The hope is that by the end of 1989, the LEP will have formed something like 100,000 Z degree particles.

With that many particles to work with, it should be possible to tell how massive each particle is with greater accuracy than has been possible until now. Then, too, it should be possible to tell how long the Z degree particle lasts before it breaks down. So far we know that it lasts about a millionth of a billionth of a second, but scientists need a figure that is more exact than that.

If the properties of the Z degree particle are known with proper accuracy, scientists believe they will be able to deduce just how many leptons and quarks there can possibly be. They rather suspect that the answer will be that there are only twelve leptons and twelve quarks possible and that all but the top-quark have been discovered.

However, even if that were so, there is still the possibility that there may be particles that are neither leptons, quarks, nor bosons but fit into different categories altogether. It may well be that the universe is far more complicated than we know.

Taming Antimatter

It may well be that there is nothing so exciting in science as the ability to make certain theoretical calculations and decide that something must exist that no one has ever observed—and then have someone observe it. This happened to a physicist named Paul Dirac.

In 1928, he was calculating the properties of electrons according to the newly worked out equations of quantum mechanics. It seemed to him that the equations showed that there ought to be *two* kinds of electrons, one the opposite of the other. Ordinary electrons have a negative electric charge, but the other kind (*antielectrons*) should have a positive electric charge yet be identical in all other ways.

No such positively charged electron had ever been observed, so few took Dirac seriously at the time. But in 1932, the physicist Carl Anderson, who was studying cosmic rays, detected the track of a particle that was created when cosmic rays smashed into the atoms of the atmosphere. The particle left a track exactly like that of an electron, but it curved in the wrong direction. This meant it had a positive charge, not a negative one. It was an antielectron or, as Anderson called it after its charge, a "positron."

Dirac promptly received a Nobel Prize in 1933 and Anderson received one in 1936.

But Dirac's equations applied to various particles. If the electron had an opposite, protons, neutrons, and many other particles must have opposites, too. All these antiparticles ought to be able to get together to form atoms that are opposite in character to ordinary atoms. These opposite atoms would make up "antimatter."

Protons are about eighteen hundred times as massive as electrons, so eighteen hundred times more energy is used to form an antiproton. Waiting for a cosmic ray energetic enough to form an antiproton would take a long time, but scientists did not have to wait. They were building larger and more powerful particle accelerators, and in the 1950s some were powerful enough to produce energies that could form antiprotons. Two scientists, Emilio Segrè and Owen Chamberlain, managed the task in 1955 and received a Nobel Prize for it in 1959.

There seems to be very little antimatter in the universe naturally, but scientists have managed to create some in the laboratory. They were able to do so chiefly because, by simple mathematical deduction, Dirac had pointed out that the task was possible.

It was difficult to study such antimatter in detail, however, because the individual antiparticles, when formed, were extremely energetic and were moving at enormous speeds. What's more, each antiparticle was bound to encounter a normal particle of its own kind in a tiny fraction of a second, since everything about them was made up of trillions of normal particles. When this happened, antiparticle and particle canceled out; both disappeared and turned into energy. This is called *mutual annihilation*.

The problem arose, then, of taming antimatter; of slowing down the antiparticles and keeping them away from the trillions of ordinary particles to avert their instant destruction. In this way, scientists would have a chance to study them in detail.

European physicists in Geneva, Switzerland, the site of Europe's most powerful instruments for studying particles, now seem to be accomplishing this. Particle accelerators have continued to grow more powerful since the 1950s. The accelerator in Geneva is one of the most powerful in existence; it can form antiprotons in huge numbers.

These antiprotons are made to pass through beryllium metal. About half of them manage to interact with protons in the metal and are lost. The rest bounce off electrons (with which they do not interact) in the outer rings of the atoms and emerge with most of their energy lost.

They emerge into a trap that contains a vacuum so that there are very few particles with which they can collide and undergo annihilation. The trap is also kept at near absolute zero, further draining the energy from the antiprotons. What's more, the trap contains a magnetic field that keeps the antiprotons bouncing back and forth and never allows them to hit the walls, where they would be annihilated. In this way, antiprotons can be kept in existence for up to ten minutes, and probably for much longer periods still.

Now scientists may have a chance to measure the mass of an antiproton exactly. Dirac's theory predicts that the antiproton's mass ought to be exactly the same as that of the proton. This is what scientists expect to find. If they do, then Dirac's theory is supported more strongly than ever.

However, the discovery of any small deviation might be as exciting as it was puzzling. It would mean that the theory would have to be modified and extended. And this might open the door to brand-new, still deeper insights into the nature of the universe.

Improving on the Diamond

Diamonds are the quintessential gems: beautiful, sparkling, rare, and expensive. Yet we seem to be on the threshold of making diamonds common, cheap, and very, very useful.

Diamond is pure carbon and carbon is one of the cheapest substances there is. Coal is carbon, for instance, and so is the graphite we use in pencils.

But if coal and diamond are both carbon, what makes them so different?

It is entirely a question of how the carbon atoms are arranged. In every form of carbon except the diamond, the carbon atoms are arranged loosely. In diamond, however, the carbon atoms are arranged very compactly. Every carbon atom in diamond is tightly surrounded by four other carbon atoms. Carbon atoms are so small and, when tightly arranged, hold together so firmly that diamond is the hardest substance known.

The trick is, of course, to force the carbon atoms into that tight and compact arrangement. To do it, you have to heat ordinary carbon to a very high temperature to enable the carbon atoms to move about more or less freely. The hot carbon is then put under huge pressure in order to squeeze the atoms together tightly. The combination of high temperature and high pressure is hard to attain, and it wasn't until 1955 that scientists at General Electric were able to convert ordinary carbon into small diamonds.

Is there any way of producing diamonds at low temperatures and pressures? One would suspect there isn't, but

Soviet chemists have been experimenting for years with a novel and very imaginative technique.

The trick is to produce a gas that contains single carbon atoms and allow these to layer themselves on to some other substance. For instance, you can begin with methane, a very common gas. Each molecule of methane contains a carbon atom attached to four hydrogen atoms.

If the methane is heated enough, the molecule is broken up into a mixture of carbon atoms and hydrogen atoms. If the vapor is then passed over a sheet of glass, for instance, the loose carbon atoms (which have a strong tendency to hook on to other atoms) will attach themselves to the atoms on the glass surface. There will be an invisible layer of carbon on the glass that would be only one atom thick.

If, however, the heated methane vapor continues to bathe the glass, additional carbon atoms attach themselves to the carbon atoms already present to form a layer several atoms thick. After years of experimentation, the Soviet chemists were gratified (and perhaps a little surprised) to find that the carbon atoms in the thicker layers took on the compact diamond arrangement.

In other words, the glass was not merely coated with carbon: it was coated with a diamond film. And only high temperature was needed to obtain the carbon-containing vapor. No high pressure was necessary.

Imagine spectacles, or sunglasses, with a diamond film. The film would be perfectly transparent and unnoticeable, but the surface of the glass would have the properties of the surface of a diamond. It couldn't be scratched, except by another diamond.

If the process is made sufficiently foolproof and routine, it is perfectly conceivable to imagine that all high-quality glass would be made with diamond film. Such "diamondized" glass, almost as cheap as ordinary glass, would be immune to scratching and scuffing.

What's more, diamond films could be formed on surfaces other than glass. Diamondized knives and razor blades would never lose their sharpness under ordinary use. Diamondized bearings and machine tools would last just about forever. And because diamond is waterproof and virtually untouched by chemicals, diamondized materials would be immune to rust and corrosion.

Diamond also is an electrical insulator and an excellent conductor of heat. This means that electronic devices can be diamondized to good effect. Electronic equipment would in this way be less affected by stray electric fields and would not accumulate heat.

Diamond can also be made into semiconductors by the proper addition of traces of boron or phosphorus. Such semiconductors would be resistant to radiation and transparent to ultraviolet light, and their electrons would move about much faster than those of other semiconductors. It is possible that by proper diamondizing, enormous advances can be made in computer technology.

More surprising still is the word coming out of the Soviet Union (which still leads the world in this technique) that new ways of making diamond films produce some sort of distorted arrangement of the carbon atoms that seems to make a film that is even harder than ordinary diamond. They are not sure what the distortion is or why it should make diamond harder, but if the initial reports are borne out, we can only guess at what more diamondizing might do.

Cold, Cold Fusion

Fusion power might well be one of humanity's great boons. The fuel isn't the moderately rare uranium or thorium, as it is in fission. Instead it is deuterium ("heavy hydrogen"), and the ocean is loaded with countless tons of it. Fission can last us for thousands of years, but fusion can last us for billions.

What's more, fusion produces far less in the way of radioactive ash, and it doesn't require a large "critical mass" to work with, as fission does. With a large critical mass, there might be a runaway meltdown. In the case of fusion, you can work with microscopic bits of deuterium at a time. Even if it gets away from you, you just get a comparatively small pop and that's it.

Fusion is a more copious energy source, and probably a much safer one. If we can get controlled and practical fusion power, our energy problems will be over—forever!

But there's a catch. (Isn't there always?) We've been trying to attain it for many years, and we haven't succeeded yet. The trouble is that in order to get nuclear fusion we have to smash one atomic nucleus into another. Atomic nuclei, however, all carry positive electric charges and positive charges repel each other.

This means that when you try to push hydrogen nuclei together, they labor with might and main to avoid each other. In order to have it *our* way, and not theirs, they have to be pushed together with a mighty shove. The way to do this is to heat the hydrogen so that the nuclei move so rapidly (the higher the temperature, the faster the motion) they don't have time to dodge. And it's not just a little bit of temperature that's needed. It's tens of millions of degrees.

Fusion takes place at the center of the Sun, where the temperature is 15 million degrees C (27 million degrees F). The center of the Sun is under the pressure of weight from the outer layers of the Sun, which also force the nuclei together. The combination of temperature and pressure does it.

There's no way we can ever produce the pressure at the center of the Sun here on Earth, so we just have to raise the temperature higher still to make up for it. We may have to go up to hundreds of millions of degrees. We have spent more than thirty-five years trying to get the temperature high enough and, so far, we haven't quite managed it.

But is there any way we can get fusion to work at low temperature? Is there any way we can get cold, cold fusion? There is a possibility we can.

At low temperatures the nucleus of each hydrogen atom has an electron in the outer ring that serves to shield the nucleus. With electrons present, the nuclei can never even approach each other, let alone smash into each other.

However, we're talking about ordinary electrons. There is another particle called a *muon*, which is like an electron in every respect that can be measured except one. The exception is that it is more massive than an electron: 207 times as massive.

Scientists don't know why it should exist, or why it should be so much more massive than an electron when it is exactly like it in every other way, but there it is.

An electron can balance the hydrogen nucleus, which always contains just one proton, to form an ordinary atom of hydrogen. Therefore, a muon can do so, too. Why not? It's just a heavy electron. In doing so, it forms a *muonic atom*.

But the muon, being 207 times heavier than an electron, circles the nucleus 207 times closer than the electron does. A muonic atom is hardly bigger than the tiny nucleus itself. In fact, under certain conditions, a muon will circle *two* hydrogen nuclei and force them very close together even at ordinary room temperature.

This will be particularly useful if one of the two hydrogen nuclei is deuterium and the other tritium (an even heavier form of hydrogen). Deuterium and tritium will fuse much more easily than two deuteriums, and if they are held together by the muons, no temperature higher than the ordinary one about us will be required.

What's more, after the fusion takes place, the muon leaves and then circles another pair of nuclei (deuterium and tritium). One muon might bring about the fusion of 150 pairs of nuclei on the average.

There are the usual catches, of course. Tritium is radioactive and exists in nature only in traces. It would have to be manufactured, and this is not an easy job. Muons are an even more ticklish proposition. Tritium, once formed, will last for twelve years on the average before breaking down. Muons, however, will only last for one two-millionth of a second. They will have to be constantly manufactured. Finally, even fifty fusions for every muon isn't enough. Ways of producing more will have to be found. They're working on that at the Rutherford Laboratory in Oxford, England.

Even if we manage to get things hot enough for ordinary fusion, it will pay to find a cooler pathway, if we can. It will, in the long run, be easier to handle and much cheaper.

Tritium— Why It's Crucial

We hear a great deal these days of something called "tritium."

We hear that we are in danger of running out of it because of the closure of plants that manufacture it, and that if we do run out we will not be able to explode our hydrogen bombs at need. We will be engaging in involuntary unilateral nuclear disarmament.

But what is tritium and why is it necessary? Isn't the stuff that operates a hydrogen bomb hydrogen? Isn't that why it's called a hydrogen bomb?

Yes, but tritium *is* a form of hydrogen. Hydrogen has three forms. The atoms of ordinary hydrogen consist of a tiny nucleus made up of one proton, with an electron circling it. All hydrogen atoms have exactly one proton in the nucleus, but some have one or two neutrons in addition. A neutron is as heavy as a proton but does not affect the chemical nature of the atom.

Thus, an ordinary hydrogen atom, with just a proton in the nucleus, is *hydrogen 1*. A hydrogen with two particles in its nucleus, a proton and a neutron, is twice as heavy, so it is *hydrogen 2*. A hydrogen with a proton and two neutrons in the nucleus is three times as heavy, so it is *hydrogen 3*. Hydrogen 2 is sometimes called "deuterium" from a Greek word meaning "second" because it is the second of the three hydrogens. Similarly, hydrogen 3 is called "tritium" from the Greek word meaning "third."

Hydrogen can be made to undergo "fusion." Its small

atoms can be mashed together under enormous heat and pressure to form larger atoms, releasing vast energies in the process.

Hydrogen 1 fuses with great difficulty. It does so in the Sun, but we can't duplicate the extreme conditions at the Sun's center here on Earth. Hydrogen 2 fuses more easily, and hydrogen 3 fuses most easily of all.

Therefore, if we want a hydrogen bomb, we would want to use hydrogen 2, rather than hydrogen 1, and hydrogen 3 (tritium), most of all—but there's a catch. Out of every 100,000 atoms of hydrogen that exist on Earth, 99,985 are hydrogen 1 and 15 are hydrogen 2. This, in itself, isn't too bad. There's so much hydrogen in the oceans and hydrogen 2 is so easy to separate that we can have tons of it, if we wish. However, hydrogen 2 by itself is still not easy enough to fuse for our purposes. We need at least some hydrogen 3, and it exists in nature only in insignificant traces. It simply cannot be obtained in useful amounts.

Why is that? Hydrogen 1 and hydrogen 2 are stable. They will retain their identity for aeons. Hydrogen 3, however, is radioactive. It breaks down into helium 3 (virtually useless in fusion reactions) at such a rate that in twelve and a half years, half of any amount of tritium has decayed. Any tritium that once existed on Earth in times past is therefore long gone. The only reason even traces exist today is that some atoms of tritium are constantly being formed in the atmosphere by cosmic rays.

However, scientists learned how to make tritium themselves by means of certain nuclear reactions, and huge plants were built in which those nuclear reactions could be carried out. Not enough tritium is formed to make hydrogen bombs of tritium only, but enough can be formed so that a little tritium can be added to deuterium to act as a "detonator," to get fusion started. The heat of the initial fusion will keep things going in the rest of the deuterium.

But the tritium-making plants must remain in operation for as long as we want to have hydrogen bombs, for the tritium they make steadily decays and there's nothing we can do about that. We must form tritium at the same rate that it decays so that we always have enough for our hydrogen bomb arsenal.

The trouble is that the tritium-making plants are old, obsolete, and unsafe; they leak radioactivity into the environment; and they form radioactive wastes that have been disposed of carelessly. For years, this has been kept secret and nothing important was done about it, as a matter of "national security."

The secret has now leaked out and people naturally object to being exposed to radiation poisoning, to cancer, birth defects, and death, even for the sake of national security. The tritium-making plants have therefore been shut down. To repair them and make them even minimally safe will cost many billions of dollars and take many years of time. To build new and better plants may take even more money and more years. Meanwhile the tritium we have is slowly breaking down, and that's the situation we're in.

Why weren't the plants kept in repair—modernized, replaced bit by bit—in the course of time? I presume because the government had other things it would rather spend the money on and it could always hide the flaws and dangers of the situation behind the smoke screen of national security.

It's depressing. How many other flaws, how many other causes of ruin, does our government hide with the cry of national security?

Forever Gone

In 1960, I wrote an essay warning that irreplaceable helium was being wasted; that, once wasted, it would be forever gone; and that someday we'd be sorry. These many years later we are still wasting helium at an enormous rate. Most of the helium we produce is simply allowed to escape into the atmosphere. Once there, it cannot be regained.

Why not? After all, are not the various elements (of which helium is one) immortal, except for a very few radioactive ones? We may use aluminum and other metals, but they can never really be used up, for we can always recycle the metals and get them back, albeit at the cost of energy.

This would be true of helium, too. Every million pounds of atmosphere contain three-quarters of a pound of helium. If we wanted to spend the energy, we could extract the helium from the air about us. Of course, helium obtained in this way would be very expensive.

But there are two gases that are so light that Earth's gravity is insufficient to retain them. Hydrogen and helium slowly but inexorably leak out of our atmosphere into space and are gone. With two out of every three atoms in our vast ocean being hydrogen, at the slow rate of leakage, it will remain in large quantities as long as the Earth is in its present state. However, helium is a very rare substance and, in its case, the atmospheric leak is serious.

Why is helium still present? Why hasn't it long since leaked away? Its origins help explain why it is still around. Helium is formed very slowly through the breakdown of radioactive uranium and thorium atoms. Over the billions of years of Earth's existence, it has accumulated in the rocks and, in particular, has mixed with pockets of natural gas. We are using this accumulated helium at billions of times the rate that it can be replaced by the further breaking down of uranium and thorium.

There are a few natural-gas wells in Texas and Wyoming that are particularly rich in helium; in fact, they produce more than 90 percent of the world's supply. However, these wells are mined primarily for the natural gas, and natural gas is wanted in such quantities that more helium is produced than can be used. The wisest move would be to separate the helium from the natural gas and store it for future use, but this would cost money. For this reason, about three-quarters of the helium obtained from these wells is simply allowed to bubble off into the air and—good-bye.

Why do we need helium? For one thing, it is the second-lightest gas, after hydrogen. Helium is used in balloons. Hydrogen would be even better, but hydrogen has a great tendency to burn (remember the *Hindenburg*?). Helium is absolutely nonflammable and is perfectly safe to use.

Helium also is the gas that is least soluble in water. It is used to replace nitrogen when air must be breathed at high pressure; it reduces the danger of the "bends," a painful and life-threatening condition.

Helium does not react chemically with any other elements. Therefore, it is used in welding as a gas that surrounds the hot flame. The material being welded will not react with the helium, as it would with air, and the weld is more likely to be perfect.

However, consider this: at 14 degrees above absolute zero, everything, with one exception, is frozen solid. Oxygen, nitrogen, and hydrogen are all solid. Only helium is still a gas.

Helium does not become a liquid until 4 degrees above absolute zero and remains so right down to absolute zero. Use of liquid helium is currently the best way to maintain the low temperatures needed for superconductivity. True, we have now discovered materials that are superconductive at much higher temperatures, but we don't know yet when they can be put to practical use. Right now, we must continue to depend on liquid helium. The superconducting supercollider being planned will need millions of cubic feet of helium every year to keep working.

Besides, even if we could make do with higher-temperature superconductivity, there are other phenomena that must be studied at the very low temperatures of liquid helium, and there is no conceivable substitute for it. We may never understand the whys and hows of the universe if we suddenly find investigation of those low temperatures preempted forever.

But is there nowhere else we can get helium if Earth's supply is recklessly allowed to vanish? Perhaps we could find helium somewhere out in space. The nearest source of copious helium (more than we can ever use) is the Sun, but how likely is it that we can collect helium from the Sun? The next nearest source is the giant planet Jupiter, which is more approachable but only marginally. I can imagine automated scoops skimming by Jupiter to take up helium from its upper atmosphere, but clearly we won't develop the necessary technology for that for a long time, if ever.

The Simpler Shape

Three chemists, two Americans (Donald J. Cram and Charles J. Pedersen) and a Frenchman (Jean-Marie Lehn), shared a Nobel Prize in 1987 for essentially simplifying a shape.

The shape is that of an enzyme molecule. Every living cell has thousands of different enzymes, each capable of bringing about some chemical reaction. In the absence of the enzyme, the chemical reaction would proceed only very slowly or not at all. With the enzyme present, the cell is a hive of rapid, interlocking reactions, all of which together maintain the chemistry of normal life.

How do the enzymes do it?

The enzymes are protein molecules. Each protein molecule is made up of a connected chain of amino acids. There are twenty kinds of amino acids that can make up a protein chain, and these can be put together in large numbers, and in *any* arrangement.

Each amino acid is composed of a chain of three atoms, one nitrogen and two carbons (N–C–C). To the middle carbon is attached a side chain and the side chain of each kind of amino acid is different. Some side chains are small, some are large; some carry an electric charge, some do not; some, with an electric charge, are positive; some are negative.

Once the amino acids string together, they fold into a three-dimensional object, and the amino-acid side chains form a lumpy uneven surface with positive and negative electric charges scattered here and there. Every arrangement of amino acids produces a surface with its own characteristic shape, and the number of possible arrangements is inconceivable.

If you began with only one amino each of the twenty amino acids, you could line them up in more than 2.4 billion billion arrangements, each producing a molecule of a slightly different shape.

Actual protein molecules, however, consist of far more than twenty amino acids. The number of each type that is present varies from a few to dozens. The number of possible arrangements of the amino acids in a hemoglobin molecule (which carries oxygen from the lungs to all the cells of the body) is 10 to the 640th power. That's a 1 followed by 640 zeros! And only one of those arrangements will work perfectly.

How does the body get that one arrangement of hemoglobin? In every cell there are chromosomes composed of molecules of deoxyribonucleic acid (DNA) of very complex structure. Sections of these DNA molecules are genes that have the ability to produce exact replicas of themselves every time a cell divides and that contain the information that guides the formation of proteins composed of one particular arrangement of amino acids and no other.

Thousands of enzymes are produced, each in countless very similar varieties. It is because the nature and balance of enzymes are different from one life-form to another that some 10 to 20 million species of living creatures have, through the aeons, evolved on Earth; that at least 2 million still exist today; and that millions more may evolve in the future. It is because of minor differences in enzymes that each individual in a species is different, so that no two human beings are exactly alike (not even identical twins).

Enzymes work through the nature of their surfaces. A particular enzyme may have a surface onto which some certain small molecule fits very neatly. That small molecule combines with the enzyme and is held in place, enabling it to combine with other molecules and undergo a chemical change. Once the chemical change has taken place, it no longer fits the surface and is released.

Each enzyme has an active center, a portion of surface made to order for the small molecule whose chemical reaction it controls. It's the active center that does the actual work, but the enzyme must have very complex additional areas that ensure that it fits in with all the other enzymes and does its work in coordination with the entire system.

That complexity of structure makes enzyme molecules large, rickety, and easily broken down. This doesn't matter in the living cell because complex enzyme molecules are built up again as fast as they break down.

Suppose you extract the enzymes from the cells, though, and try to use them to cause certain chemical reactions to take place? The problem is that they break down rapidly and we don't have the facilities to build them up again quickly enough.

But what if you manufacture a small molecule that has the shape of the *active center* of an enzyme? It might not work as smoothly and as well as an enzyme does, but it might do the work well enough. Its simpler structure would make it much easier for chemists to re-create. What's more, a synthetic molecule with a desirable shape in its simplest form would be much more stable than an enzyme.

It is these molecules of simpler shape that the three chemists have been manufacturing. The simplified enzymes are now used in medical diagnosis. The chemists' work has been credited with laying the foundation for a field of biomedical research that has been expanding rapidly.

The Dangerous Microorganism

It has always been taken for granted that a scientist has a right to experiment as he chooses. The quest for knowledge has seemed so noble in intent, so rarefied in principle, so often useful in the consequence, that even repressive societies have usually left it alone. And yet there are potential dangers in experimentation and some kinds are regulated—and rightfully so.

For instance, it would not make sense to build a laboratory intended for the study of high explosives in a densely populated urban area. If chemical experiments yielded foul odors or poisonous gases, there would be laws to minimize the discomfort and danger these could cause.

In the same way, the use of genetically altered microorganisms (protozoa, bacteria, viruses) is viewed with a certain unease even by scientists.

It is perfectly possible these days to alter the genetic content of microorganisms to give them abilities they do not have in nature. A bacterium might be given genes that would make it possible for it to manufacture in quantity human insulin, or some other protein of value to medicine. Other bacteria might be developed that had the ability to live on oil spills or to digest otherwise indigestible waste; such novel microorganisms might well contribute to the cleansing and detoxifying of the environment.

These potential useful advances are just an example of hosts of attractive applications of the new art of "genetic engineering." And yet once an altered microorganism is released into the environment, it may well spread and become a permanent fixture impossible to eradicate. What, then, if you decide that, on the whole the altered microorganism isn't as useful as you thought—even, perhaps, harmful?

Too late! You can't squeeze the omelette back into the eggshell.

In altering a microorganism, suppose you find that you have inadvertently created a variety that has the power of producing a highly dangerous toxin within the human body. Suppose also that it happens to multiply quickly and spread from person to person by contaminating air or water. You might end up with the incredible nightmare of another Black Death that would kill hundreds of millions of people in a sudden, devastating pandemic.

This is not really likely, but suppose you merely manufacture a bacterium that induces bouts of diarrhea or something equivalent to intestinal flu. It might not kill, but the inconvenience would be severe. The fury of the public against the perpetrators would be extreme.

It is therefore in the public interest to have the government regulate such experimentation, to insist on certain safeguards, to have the possible effects of any altered microorganism closely considered by experts before the chance of releasing it into the environment is taken.

But such regulation is difficult and can easily be made to seem unreasonable. And what if the regulatory board takes exorbitant lengths of time or plays it so safe that genetic engineering is throttled?

For instance, a scientist named Gary Strobel at Montana State University has worked with a bacterium that produces an antibiotic that kills the fungus that causes Dutch elm disease, a disease that has all but destroyed the beautiful elm trees of the world. Strobel genetically altered the bacterium in order to give it the ability to produce greater quantities of the antibiotic and then injected those bacteria into fourteen elm trees to see whether it would make them immune to the disease.

By the time he had gotten around to making the necessary application to the Environmental Protection Agency (EPA), he realized that waiting for such permission would delay his experiment for a year. Therefore, he decided not to wait and performed the experiment without the necessary permission.

He reasoned that the bacteria would be inside the tree, could not make contact with people, and would have no effect on people if they did; that the aim was a good one and the chance for harm nonexistent.

The EPA was forced to decide what to do. To do nothing at all might make it appear that the regulatory agency was a paper tiger, utterly toothless, that might safely be disregarded by anyone. The alternative was to punish the experimenter with a warning or a fine or even a prison sentence. Indeed, the punishment could have extended beyond him. The university at which he works could have lost its federal funds.

Severe punishment is capable of making scientists indignant, but it can also caution against a repetition of Dr. Strobel's defiance. In the end, Dr. Strobel was forced to cut down and destroy his fourteen elm trees to prevent trouble for himself and his university.

The Telltale Flash

To diagnose a bacterial infection the particular bacteria present must be identified. You must take a sample of a bodily fluid, grow the bacteria it may contain, and study them. After days have passed, you may be able to tell which bacteria caused the infection and to decide what to do about it. Happily, it might be possible in a few years to do it all in ten minutes and find out the result through a flash of light, a very special light flash that involves no heat.

Ordinarily, we associate light with heat. If something is hot enough it will radiate light. The usual way of bringing this about is to burn something. The combination of fuel and oxygen will yield heat and light. Or we may run an electric current through a thin, resistant filament, which will heat up and yield light. Such *incandescent light* is inefficient, because most of the energy is lost as heat.

It is only in recent times that we learned practical ways of shining ultraviolet light on certain powders that absorb the ultraviolet and then give off the energy gained as visible light, with very little heat. This *fluorescent light* is more efficient than incandescent light.

But in this area, nature has beaten us to the punch by millions of years. There are forms of life, mostly bacterial and mostly living in the sea, that radiate light called *bioluminescent*, without any heat at all. This phenomenon is rare on land but we do have fireflies, which periodically flash a dim light from their abdomens. They do this primarily as a signal that enables them to find each other for mating purposes. This interests not only biologists but also chemists who want to learn how the light might be produced.

The firefly contains a rather unusual compound that has been named *luciferin* (Latin for "light-bearing"). Ordinarily, it just sits there, but in the presence of an enzyme called *luciferase* (also present in fireflies) it can easily react with a molecule of adenosine triphosphate (ATP). This substance is present in all living cells without exception, from the cells of bacteria to those of human beings.

ATP is a "high-energy" compound, and its function in cells is to deliver bits of energy to places where it is needed (which is why it has to be present everywhere and why we can't live without it). When ATP transfers a bit of energy to the luciferin molecule, that molecule is changed to a slightly different molecule, called "oxyluciferin." Oxyluciferin, having received energy in this way, is rather unstable and has a strong tendency to give up that extra energy and return to the more stable, less energetic luciferin, almost immediately.

In living cells, all sorts of energetic molecules are formed, and they usually give off their energy by passing it on to some other molecule, which, in turn, passes it on to still another, and so on. Such juggling of energy from molecule to molecule makes possible all the complex reactions that characterize the living state. In the process a certain amount of heat is produced that, in the case of birds and mammals, keeps the body temperature fairly high and constant.

Oxyluciferin, however, is exceptional. Its energy is not passed along to other molecules but is given off as a tiny flash of light, without heat. This unusual action is not limited to the interior of the firefly's cells. Luciferin and luciferase can easily be extracted from the firefly. If a bit of ATP is added to a solution of these materials, luciferin changes to oxyluciferin, which breaks down to luciferin again so that the entire solution begins to glow.

Well, now, suppose you dip a small stick into, let us say, a urine sample, and then dip the stick into a mixture of luciferin and luciferase. Normal urine has little or no ATP, meaning that nothing much will happen to the mixture of luciferin and luciferase. But what if there is an infection of the urinary tract that causes bacterial cells (and, therefore, ATP) to be present in the urine? In that case, the luciferin-luciferase mixture will flash its light.

To make the test more delicate, you treat the urine to destroy any ATP it may possess other than that which is present in bacterial cells. You also don't depend on the eye to see the light but make use of special devices that can not only detect dim light but can measure its brightness accurately. The brighter the light, the higher the concentration of bacteria.

Naturally, all bacteria will supply ATP and produce the light, but it is possible to use antibiotics that will kill off certain bacteria that do not interest you, leaving behind those that do: making the luciferin-luciferase test-specific.

In this way, for instance, all sorts of food and drink can be tested quickly and accurately for bacterial contamination. For instance, eggs and chickens could be much more easily tested for salmonella, infections from which have been particularly bad and occasionally fatal.

There's still a way to go to make the tests practical and accurate, but the principle is there.

The Genome Project

James Dewey Watson is one of America's greatest biochemists. He helped work out the structure of the double helix of DNA, the basic component of life, in 1953. For this he received a share of the Nobel Prize for physiology and medicine in 1962. And, in autumn 1988, he was chosen to head the genome project.

What is the genome project? In fact, what is the genome?

To begin with, the infinite complexity of life rests on the fact that in every cell there are many thousands of chemical reactions, all taking place simultaneously and each affecting all the others. In no two species, or, for that matter, two individuals of the same species, is the chemistry precisely alike. It is even different in different cells of the same individual. The nature of each organism, each speck of life, is determined by the interlocking chemical reactions.

Each chemical reaction is controlled by a different enzyme molecule, a large complex molecule composed of dozens or hundreds of smaller units, the twenty kinds of amino acids, which are arranged in a chain. If a single amino acid is out of place or is slightly modified in place, a particular enzyme's capacity to do its work may be completely altered.

Each enzyme is produced according to the directions contained in the genes within the chromosomes that are located in the nucleus of the cell. Each gene consists of a long string of nucleic acid molecules, the structures of which were determined by Watson and his co-worker, Francis Crick. The nucleic acid consists of thousands of units called "nucleotides," of four different varieties, all arranged in a double helix, which looks like two bedsprings pushed together. The four different nucleotides are usually known by the initial letters of their names: A, C, G, and T.

If we could determine the exact order of all the nucleotides in the nucleic acids of the human chromosomes, we could have a series of letters—AACGTGTCGAA . . . and so on—that would make up the "human genome." If we take these letters three at a time, each "triplet" represents a particular amino acid. The order of the triplets determines the order of the amino acids in the enzyme, and therefore the enzymes' structure.

Knowing the genome, we would take a giant step toward grasping the scheme of the human being in detail. This, in turn, would enable us to understand all the chemical reactions and, perhaps, how they affect each other.

It's not a simple project. In the human being, there are perhaps 50,000 enzymes controlling 50,000 chemical reactions. The nucleic acid molecules that contain the information that leads to the formation of these enzymes are made up of 6 billion nucleotides. If these nucleotides are represented by letters in the correct order, this would represent a billion words, or, roughly, the equivalent of 360 books, each the size of a volume of the *Encyclopaedia Britannica*. That's a lot of information, but, then, a complete knowledge of the human body requires a lot of information.

Even today, only about a thousandth of the nucleotides have been placed in order, a few here and a few there, but the methods for determining the order of the nucleotides are rapidly being improved and automated. It may be that in a few decades, the human genome will be worked out.

Even that will not be enough. It will give us the *standard human genome*, one that will be found in a normal, healthy human being. However, many people are born with defective genes of one kind or another and may have serious inborn and inheritable metabolic abnormalities as a result. There are at least four thousand of such identified abnormalities known to be encoded in the genome at birth, and it is important that we have methods for the detection of these abnormalities in the genome as early in life as possible.

What's more, even among normal, healthy human beings, genes exist in varieties that may not imply serious abnormalities but that still introduce individual characteristics. Genes, in different varieties, lend different colors to the eyes and hair, different shapes to noses and chins, and varied appearances in height and structure. It would be useful to identify the important varieties so that every individual can be assigned his own genome and his own nucleic acid "fingerprint."

This will not necessarily mean that every person will have to carry his entire genome in 360 giant volumes in his personal computer, of course, merely a record of the major variations.

Eventually, perhaps, scientists will determine the genomes for every one of the 2 million species of plants, animals, and microscopic organisms now alive. This might give us a keener understanding of relationships among living things and a more detailed look at the course of biological evolution.

We might even work out, in theory, new kinds of genomes that no species possesses and learn some of the might-have-beens of biology. It is indeed an enormous project that Dr. Watson is undertaking, and he is eminently well qualified to head it.

First Look at the DNA Molecule

In January 1989 scientists got their first direct look at a very important molecule.

Scientists have been looking at objects that were too small to see with the naked eye for nearly four hundred years. The trick at first was to use lenses that forced light to bend and in this way focus and enlarge the image of the object that reflects the light. The device used to do this was the microscope.

As time went on, microscopes were improved until finally they could magnify objects one thousand times. At that point, scientists ran up against a physical barrier. Light consists of waves. These waves are tiny, but the objects under the microscopes were tiny, too. If the objects were tiny enough, they were smaller than the light waves being used to view them. The light waves then tended to skip over them so the objects could not be seen.

To get around this, you might use shorter light waves such as ultraviolet. For a while, therefore, scientists used "ultramicroscopes," but these represented only a small improvement. Still shorter waves could not be focused properly.

But in 1923 a French scientist, Louis de Broglie, pointed out that subatomic particles ought to exist in wave form, too. In 1925, an American scientist, Clinton J. Davisson, was able to detect such waves produced by electrons. These electron waves were much shorter than ordinary light waves. They were, in fact, about the size of X-ray waves. But where

X-ray waves could be extremely difficult to focus, electrons and their waves could be focused easily by magnetic fields.

The first device used to focus electron waves and enlarge the images of objects in that way was constructed in 1932 by a German scientist, Ernest Ruska. This was an "electron microscope." It was crude at first, but over the years it was refined and improved, until it could enlarge objects 300,000 times.

At first, in such instruments, electrons had to pass through an object to produce an enlarged image. Scientists had to work with very thin slices of material. But then, ways were found to produce a very thin, sharp beam of electrons and play that beam over the surface of an object. The beam "felt" its way, so to speak, across the surface, scanning it and producing an enlarged image. This was a *scanning electron microscope*.

Now, a still newer version produces the electrons by means of what scientists call a "tunneling effect" and we have a "scanning tunneling electron microscope" that has reached new heights of magnification. It can magnify one million times and was used to take a first-ever look at a molecule of deoxyribonucleic acid (DNA) by Miguel B. Salmeron and others at the Lawrence Livermore Laboratory in California early in 1988.

DNA is important because it carries the blueprint of life. Every living cell contains a set of DNA molecules that are constantly replicating themselves, passing the newly created molecules on to daughter cells. Such sets also exist in sperm and egg cells, so that they can be passed on from parents to children. Every living species has its own set, and there are also tiny differences in the sets of different individuals of the same species.

The importance of DNA was first recognized in 1944, and scientists labored to find out how these molecules could produce other molecules exactly like themselves.

In 1953, a British scientist, Francis H. C. Crick, and his American co-worker, James D. Watson, worked it out. X rays, in passing through molecules, tend to bounce to the side. Photographs of such X rays produce dots where the X rays bounce, and from such "X-ray diffraction patterns" it is possible to deduce the shape of the molecule. After a long time and careful work, it finally turned out that the DNA molecule consisted of two complex strands of atoms twined in a double helix (the shape of bedsprings). Each strand had a complicated shape, and the two strands fitted each other exactly.

When DNA forms another molecule like itself, the two strands unwind and each one picks up small groups of atoms from the cell fluid and puts them together into a new strand that fits exactly upon the original one. Each strand serves as a model to form a new partner for itself. In the end, then, each DNA molecule produces two DNA molecules exactly alike.

This discovery brought Watson and Crick the Nobel Prize in 1962, and their work was considered a triumph of subtle scientific deduction. They described the double helix of the DNA molecule and its workings in precise detail, even though it was actually much too small to be seen.

But thirty-six years after the work of Watson and Crick, pictures of a DNA molecule have been taken with a scanning tunneling electron microscope—and no subtle reasoning is required. There is the double helix, visible to the eye. The molecule coils as it is supposed to. From the image, it is possible to work out the distance between successive coils. It is about one-five-millionth of an inch.

Salmeron and his group are planning to refine the method further to see whether they can see even finer details of the strands. They will try to obtain images of other molecules, too.

The Head of a Pin

We have all heard that in the Middle Ages scholars would debate the question of how many angels could dance on the head of a pin. I'm afraid that I've never heard the outcome of the debate. Nowadays, though, scientists can ask themselves a similar question and come up with a rather surprising answer.

If we ask about angels, we are asking about supernatural beings, and that means they are not bound by the laws of nature. Each angel can therefore, if it wishes, take up no space at all. If we argue in that fashion, then an infinite number of angels can dance on the head of a pin. I don't know whether that was the answer the medieval scholars accepted, but even if they did, there was no way in which they could demonstrate the truth of the hypothesis.

Scientists, however, are bound by the laws of nature, so that if they ask, for instance, how many words can be inscribed on the head of a pin, they know there is no way of inscribing an infinite number. Each letter takes up room and there isn't much room on the head of a pin, so the belief is that you can't inscribe many words. If we look at a pin, it seems to us that inscribing even one word would be an immense task.

Of course, we sometimes read about people who manage to inscribe the Lord's Prayer on a pinhead. I imagine that they must use a strong magnifying glass and a very sharp stylus, and that they must have a very steady hand. They inscribe the very tiny letters so that someone else with a strong magnifying glass can read, "Our Father, which art in Heaven . . ." and be amazed.

But what can scientists do now? Well, they can use not a stylus but a sharply focused beam of electrons. Researchers at the University of Liverpool report that they use a beam that is so sharply focused that it bores a line that is only two atoms wide. Such a line is so narrow that they could draw one million lines side by side in the width of an ordinary pencil line. Naturally this can't possibly be done by hand. The tiniest tremor would blur those lines hopelessly. The electron beam has to be under the control of a computer.

Now let's consider a pinhead. A reasonable pinhead might be a millimeter (one-fiftieth of an inch) wide. The atoms that make it up are so tiny (less than a hundred-millionth of an inch across) that there are about 4 trillion (4,000,000,000,000) atoms on that pinhead.

Suppose we imagine all those atoms grouped into squares, each of which contains 12 atoms on each side, or 144 atoms altogether. There would be 28 billion (28,000,000,000) of these squares on the pinhead. In each of these squares a letter could be incised, and some squares could be left blank, serving as a space between letters.

Since there is an average of six letters per word (counting the space between the word and the next word as a blank letter), the 28 billion squares can be filled with 4.7 billion (4,700,000,000) words.

That is a lot of words. Studying my copy of the *Encyclopaedia Britannica,* I conclude that it contains about 50 million (50,000,000) words. This means that we can squeeze onto the pinhead the complete *Encyclopaedia Britannica* and, in the process, use up just a little more than one percent of the space. You could make each letter ten times as tall and ten times as broad as I have suggested and still squeeze the *Encyclopaedia Britannica* onto the pinhead.

The people at the University of Liverpool have shown

this can indeed be done by copying one page of the encyclopedia on a pinhead in a small enough scale that the entire encyclopedia would have fitted had they continued. (Of course, there isn't much point in placing the encyclopedia on a pinhead, other than to show how delicate the technique is.)

The electron beam can be used, more profitably, as an analytic tool, telling scientists just what kind of elements are present in particular places. Thus, one problem they can be applied to is that of superconductive materials that vary from batch to batch in accordance with tiny changes in proportions of the elements present. The electron beam can be focused at boundaries between layers of crystals in the superconductor, and this will cause the atoms to give off X rays. The exact wavelength of the X rays depends on the nature of the atoms present, and, in this way, the electron beam can identify the different elements present and the precise quantity of each.

This sort of analysis may teach us a great deal about superconductivity and help us make use of it in various technologies in ways that could change our society completely.

It seems to me, too, that something capable of scoring material so finely might be useful in producing tiny circuits on computer chips. Billions of circuits could easily be made to fit on a single chip a quarter-inch to the side, and if superconductive materials are used, no disabling heat will be developed. Truly complex thinking machines might be built, and that might mean artificial intelligence that is as capable as our own (or more capable, perhaps). Whether we would want to do that is, of course, another question.

Our Biological Clock

For four months early in 1989, a young woman named Stefania Follini remained voluntarily underground. She stayed in a Plexiglas module that was twenty feet long and twelve feet wide, about the size of a comfortably sized living room, and this living place was thirty feet underground near Carlsbad, New Mexico. There was no sunlight, no clock, no way of telling the time at all. She did her work alone under conditions that were comfortable but timeless.

The question was: What would all this do to her *biological clock*, her innate sense of time? The answer was that the biological clock went to pieces during a prolonged period without external clues.

We all have a biological clock, every one of us: a clock that keeps our body's functions to a set of various rhythms. In short, we've all got natural rhythm.

When it is mealtime, we feel hungry. When sleep time approaches, we feel sleepy. We don't have to look at a clock to know that it is time to sleep or time to eat.

We rouse ourselves at more or less the proper time in the morning, even when daylight has not yet arrived to wake us. (As a personal note, I am an early riser, waking at 5:00 A.M. winter and summer, in sunshine, clouds, or at night, virtually never oversleep by more than a few minutes, and don't even own an alarm clock.)

Clearly, the rhythm involved in waking and sleeping keeps one more or less in time with the Sun. Most of our best-known rhythms involve ups and downs that repeat on a daily basis. These are the *circadian rhythms* (from Latin words meaning "about a day").

There are also monthly rhythms for various forms of shore life, as the tides grow higher or lower in accordance with the relative positions of the Moon and the Sun. There are annual rhythms that govern such phenomena as the migrations of animals and birds with the changing seasons. Undoubtedly, human beings have subtle rhythms of this longer sort also, but the daily rhythm is the one that is most noticeable.

It is not just the rhythm of eating and sleeping that fluctuates daily. Moods and attitudes do, too. If you wake up at 3:00 A.M. and consider a particular problem, it may seem to you that the difficulties are insuperable. The same problem viewed at 11:00 A.M. seems rather trivial. The problem hasn't changed. Only your mood has.

From the medical standpoint, circadian rhythms may be crucial. A person's reaction to drugs or character of allergic response varies also according to a circadian rhythm, and some physicians are learning to take this into account in prescribing medication. What's more, the rhythm is not necessarily the same in everyone. There are "morning people" and "night people."

Anything that upsets the rhythm can greatly lower efficiency. People who have to switch in and out of night shifts, for instance, can have trouble reacting well to emergencies. They must face 11:00 A.M. problems with 3:00 A.M. bodies.

Long rapid travels, east or west, make you arrive at a much different local time from the one you left, and again you are out of rhythm; this is called *jet lag*. Travelers are advised to wait a while, and grow accustomed to the new rhythm, before trying to make crucial decisions.

So what happened to Miss Follini's biological clock during the four-month period of timelessness, with no external cues to keep her rhythms on track? Her time sense went completely awry. She fell into a rhythm with ups and downs at only half the normal rate. She would work for up to thirty hours at a time. She would sleep for anywhere from twenty-two to twenty-four hours. There were longer intervals between meals and she lost seventeen pounds. Her menstrual period (a more or less monthly rhythm) stopped altogether. She ended up thinking she had been underground for two months, not four, and when she emerged in May, she thought it was mid-March.

All this study of biological clocks is important from the theoretical standpoint, but it has its practical aspects too.

We can rely on external triggers to be right as long as we are on the surface of the Earth. The time will come, however, when we are out in space. On the Moon, the "day" is two weeks long, and so is the "night." On a rotating space settlement, the "day" and "night" may be each two minutes long. In an underground settlement on Earth, or in a windowless settlement in space, there may be no "day" or "night" at all.

It would be necessary, then, in all these circumstances, to manage ways of setting up an artificial day-night alternation that would last a period of twenty-four hours. After all, our bodies have evolved over the aeons to suit such a rhythm, and we ought to respect that.

III

FRONTIERS OF EARTH

The Shifting Earth

Though today's Floridians won't have to worry about bundling up soon, scientists have documented that North America is drifting slowly toward the North Pole, and some believe a dramatic shift of the Earth more than 70 million years ago may have caused the extinction of the dinosaurs.

We can tell the way this shift takes place by the position of the poles. In the last eighty years, the North Pole has moved about thirty-three feet in the direction of eastern Canada, about five inches a year. This isn't because the pole itself moves, but because the Earth's surface is shifting under it and North America is moving toward it in a slanting way.

This shift of the Earth isn't much, but if it continues at this rate and in this direction, in 10 million years, New York might find itself eight hundred miles closer to the North Pole. Of course, in all likelihood, the shifting motion is at different speeds and in different directions with time, so we can't predict easily where a particular spot on Earth's surface will be millions of years in the future—or where it was millions of years in the past.

As for the future, we might let that take care of itself. The past is another matter, however, for the course of evolution may have depended in part on where the landmasses were located with reference to the poles at different times.

One of the reasons the Earth shifts has been well understood for a quarter of a century. The Earth's crust is broken into half a dozen large plates plus a few smaller ones, and these are all moving with respect to each other (some scientists think) because of the dragging effect of slow swirls in the very hot molten rock far below the surface.

The plates can include whole continents on their surface; for example, the North American plate carries North America on its back. If this plate is moving very slowly northward, North America is moving very slowly northward with it and is approaching the North Pole.

However, it seems that these plate movements do not entirely account for the shifting positions of the continents. There is an additional shift of the entire Earth this way or that, at rates sometimes as fast as the plate movements. Possibly, for relatively short periods of time, the whole-Earth shift is even faster.

How can scientists be certain of this? A British geologist, Roy Livermore, and his associates have undertaken this task. To begin with, they measured the rate at which different plates are moving now.

Then they studied the magnetic alignment in old rocks, and that gave them the position of the magnetic north pole. The magnetic north pole is not exactly at the geographic North Pole, but over millions of years, the average positions of the two kinds of poles are much the same.

Knowing the measured plate motions and the magnetic alignments, the scientists could calculate where the North Pole was located, relative to the continents, at different times in the past. This gave them the overall plate motions for many millions of years.

But what about the overall shift of the Earth as a whole?

To determine this, Livermore and his associates studied certain *hot spots*, where the very hot rock below manages to seep its way to the surface. These hot spots don't move as the plates do, for the plates are part of the crust and the hot spots originate below the crust. Thus, as the Pacific plate moved, the hot spot formed a series of volcanic islands that make up the present state of Hawaii. Such lines of islands may change direction, and this would indicate a shift in the Earth as a whole.

From his studies, Livermore believes that over the last 90 million years, the North Pole has shifted about one-fifth of an inch a year, on the average, for a total shift of about 285 miles. However, between 70 and 100 million years ago, when the lines of hot spots curved, it would seem the total shift was about 1,000 miles. Even if the total shift was spread over the entire period of 30 million years it would be three times the rate it is now. If it occurred in a smaller interval of time, it would be faster still. This rapid shift was likely due to a movement of the Earth as a whole, rather than to plate movements, which would be expected to maintain a steadier rate.

What could cause the Earth to shift as a whole? The most likely explanation is a change in the mass distribution of Earth. During an ice age, vast quantities of water shift from the ocean to huge icecaps in the Arctic. This shift of mass northward causes the Earth to shift its position a bit.

Again, when two landmasses collide, as when India collided with Asia about 40 million years ago, much of the crust slides into the hot rock beneath and is distributed over the Earth. This shifts mass, too, and causes the Earth to twist. There may even be mass changes deep beneath the crust.

Some such mass shift may have happened 70 million years ago or so. Whatever its nature, could it have led to the extinction of the dinosaurs about 65 million years ago? For now, scientists can only wonder.

The Wobbling Earth

The Earth spins on its axis. If it were a perfect sphere in shape, perfectly symmetrical in its internal makeup, perfectly rigid, and perfectly alone in space, it would spin eternally about an immovable axis. None of this is true, however, so Earth wobbles. Three different wobbles were known prior to July 1988, when a fourth wobble was discovered and announced.

When the motions of the stars during the night are closely studied, they are found to make circles about a certain point in the sky that is just over Earth's North Pole. (The North Star is near that point but not exactly on it.) If the stars are studied year after year, this central point can be seen to shift slowly. It does so because Earth's axis shifts, and this is because the Earth is not a perfect sphere but bulges at the equator.

The Moon and Sun pull at that bulge and cause the Earth's axis to move in a slow circle. The circle is completed in about 26,000 years. This effect is called the *precession of the equinoxes*, because, as a result of the motion, the equinoxes arrive a little earlier each year than they did the year before. This is the largest wobble of Earth's axis and was discovered by the ancient Greeks.

The Earth's axis does not describe a perfect circle as it moves. The Moon's pull changes slightly with time because it is sometimes a bit closer to Earth than at other times. This produces a minor wobble in the circle of the precession, a tiny wave that repeats itself every nineteen years.

This discovery was made in 1748 by a British astronomer, James Bradley, from his careful study of the position of the stars. This slight wavy motion is called *nutation*, from the Latin word for "nodding," because the axis seems to nod slightly as it marks out the circle of the precession of the equinoxes.

But this is not all. As early as 1765, a Swiss mathematician, Leonhard Euler, predicted that Earth's poles ought to move in tiny circles over a one-year period. The motion was too small for anyone to detect at the time, but as the decades passed, telescopes and other instruments became more precise and delicate.

Finally, in 1892, an American astronomer, Seth C. Chandler, was able to study the stars so accurately that he could detect very tiny shifts in their position that could best be explained by the shifting of the position of the Earth's poles. This was called the *Chandler wobble*.

The Chandler wobble is a roughly circular movement of the poles. The circle is completed in about 430 days. It is not an exact circle but tends to be wider some years than others. It is a tiny wobble and the poles' change in position in the course of a year is only about ten yards or so. You wouldn't think this would be large enough to detect; the fact that it was detected shows how refined astronomical instruments have become.

If this were the motion that Euler predicted, it ought to fade out after a while, but it doesn't. It keeps right on going. Astronomers believe this is because the distribution of matter in the Earth changes from time to time. Usually, this is a result of a large earthquake, which shifts the balance of rocks inside the Earth—not much, but just enough to put a kink in Earth rotation that slowly shifts the pole a few yards. Naturally, the more severe the earthquake, the greater the deviation, which is why the Chandler wobble is greater some years than others.

But it doesn't take an earthquake to make the Earth wobble. Any shift in Earth's mass distribution, even a tiny one, will produce wobbles, as the British scientist Lord Kelvin predicted in 1862. The smaller the shift, of course, the smaller the wobble.

Methods for detecting changes in position of the Moon or of artificial satellites have continued to improve. Laser beams can now be bounced off such bodies, and by measuring the time it takes them to return, changes in position of as little as a couple of inches can be detected. Using such techniques, scientists from the Jet Propulsion Laboratory in Pasadena, California, and from Atmospheric and Environmental Research in Cambridge, Massachusetts, announced the existence of a fourth wobble that moves the axis in a small circle in anywhere from two weeks to a couple of months. This circle has a width of two and a half inches to two feet, so that it is only about one-thirtieth as large as the Chandler wobble.

By carefully studying satellite reports on weather data, the scientists have concluded that this short-term, very small fourth wobble is produced by the change in mass distribution when winds cause the atmosphere to slosh back and forth. Other factors might be storms causing the back-and-forth movement of water, the advance and retreat of snow cover, and so on.

It is astonishing to think that phenomena as familiar as gusts of wind, or the flowing of rivers, or the melting of snow can produce a tiny wobble on the vast and massive Earth, but they apparently do.

Those Oceanic Hot Spots

Another suggested origin of life has turned out to be unsatisfactory, and scientists are left with a puzzle they have been brooding over for more than half a century.

Unmistakable traces of bacterial cells have been found in rocks that are 3.5 billion years old, and Earth is 4.6 billion years old. Some time in the first billion years of Earth's existence, then, living things evolved out of nonliving chemicals. But how? What are the details?

The trouble is that no one was there to watch and we have no time machine in which to go back. We can only reason out the matter from what we can observe on Earth and in the universe today.

Scientists have worked out the general chemical structure of the early Earth. For instance, in its youth, Earth's atmosphere contained no oxygen; oxygen is a product of more recent life. Earth's original atmosphere was composed largely of carbon dioxide and nitrogen with perhaps some methane and ammonia. The ocean was probably full of dissolved carbon dioxide, like seltzer; or full of ammonia, like a window-cleaning compound; or both.

Energy poured down on the air and ocean from sunlight that was rich in ultraviolet light because, without oxygen, no ozone layer formed in the upper atmosphere to block the ultraviolet. There were also volcanic action to supply heat and lightning to supply electrical energy.

The energy would build up the carbon dioxide or methane in the air and ocean into more and more complicated carbon compounds until the replicating properties of life appeared.

Scientists have tried to work out the exact pathway by which this would have taken place, but none seemed completely satisfactory. We need more information. This is why there was such disappointment when it turned out there were no carbon compounds in the soil of the Moon or of Mars. If such compounds had been present, they might have represented a stage in the journey toward life and could have given us the necessary additional information.

Then, of course, it might be that our difficulties arise from the fact that we have been heading in the wrong direction altogether.

In 1977, for instance, deep-sea submarine work revealed that there are certain places in the ocean bottom where the heat of the inner layers of the Earth comes close enough to the seafloor to produce "chimneys" from which hot water, rich in minerals, curls upward into the cooler ocean. About such "hot spots," bacteria lived. These bacteria get their energy from chemical changes in the minerals that continually spew upward, particularly those containing sulfur atoms.

Small animals feed on these bacteria, and larger animals feed on these small ones. A whole community of life, the existence of which had never been suspected, depended on the energy of these hot spots instead of the energy of the Sun.

Perhaps life had formed on the ocean surface, as we have been thinking for decades, and evolutionary pressures forced some bacteria downward, adapting them to life in the hot spots. However, scientists were unable to work out a convincing rationale to explain how this might have happened.

But is it possible that it was at the hot spots that life first formed, and from the hot spots that it spread to the ocean surface? If so, this would account for scientists' inability to work out a scenario for life's origin on the ocean surface: because it did not originate there.

Certain observations seemed to support this theory. The hot spots seem to have existed from the time the ocean first formed, long before life had come into existence, and these superheated vents have offered a stable environment for billions of years. Their position at the bottom of the ocean would protect the fragile beginnings of life from the disruptive effects of strong ultraviolet light and from the disturbances of volcanic heat and meteoric bombardment that were much more common in early days. Then, too, the minerals in which the hot spots are rich are just those that are important to life.

For a while, then, there was a flash of hope that the details of the origin of life, under these new and totally different conditions, might be worked out.

In 1988, however, two scientists, S. I. Miller and J. L. Bada, reported on an exhaustive study of the conditions in the hot spots and the way these conditions might affect the developing molecules that were building up in the direction of life.

It turns out, disappointingly, that the hot spots are too hot. Their high temperatures would cause any amino acids (the basic components of proteins) that happened to be formed there to break down to simpler substances in minutes. Any sugar that formed would break down in seconds. There seems no way in which the proteins and nucleic acids, essential to life, could ever have formed under those conditions.

So the bacteria that form the basis of life in the hot spots must have come into being elsewhere. This conclusion takes scientists' search for the origin of life back to the ocean surface.

The Big Crack

The most severe earthquake in the history of the United States took place on February 7, 1812, not in California but on the Mississippi River, near where New Madrid, Missouri, now stands. It destroyed 150,000 acres of timberland, changed the course of the Mississippi in several places, drained some swamps, and formed some lakes. The trembling was felt as far away as Boston.

However, this was largely unsettled territory at the time and, as far as we know, not one person was killed and virtually no damage was done to private property, so the memory faded.

In comparison, the San Francisco earthquake of 1906 was rather small potatoes, but it struck a city. It killed five hundred people and about $60 million of property was destroyed, either directly by the earthquake or by the fire that followed. That made the 1906 earthquake the best-known incident of the sort, and the grisliest, in our history.

The boundary between the Pacific plate and the North American plate is a big crack in the crust that passes through western California from San Francisco to Los Angeles, called the *San Andreas fault.* The Pacific plate is slowly turning counterclockwise so that the western edge of California is moving northward compared with the rest of the state.

If the fault were smooth, the western edge of California would slide northward a tiny fraction of an inch each year and the motion would be just about unnoticeable, except for refined scientific measurements, and it would certainly bother nobody. This, however, is not the way it works.

The two plate edges are held together under huge pressure along a very irregular rocky line. The friction of one edge against the other is enormous, and the two are kept in place even as the Pacific plate exerts growing force to turn. (It is similar to your trying to open a jar lid that has been screwed on very tightly. You put more and more turning pressure on the lid, grunting and squeezing, until finally the friction is overcome and the lid suddenly turns.)

In the same way, the inexorable twisting of the Pacific plate puts more and more pressure on the San Andreas fault, until in one place or another it suddenly gives and moves. The San Andreas fault, in the region near San Francisco, moved twenty feet in just a few minutes in 1906. The jiggling, as one irregular edge bumps its way forward against the other, causes the monster vibrations we call an earthquake. No other type of natural catastrophe, except the impact of a large meteorite, can kill so many and destroy so much in so short a time as the vibrations of a plate edge jarring its way suddenly forward.

Once the sudden movement has stopped, the pressure on the fault is relieved. It slowly builds again, but many decades may pass before the pressure can accumulate to the point of another major earthquake.

However, even though major earthquakes are few, there are often minor adjustments at different places along faults. (The San Andreas is only one of many, though it is the most famous.) The result is a frequent drizzle of minor earthquakes that do little damage and are even beneficial, for they relieve some pressure and postpone the inevitable day of the major earthquake.

Naturally, scientists want to know as much as possible about the movements of faults, so that they can learn to predict earthquakes and allow people to evacuate the area or to protect their property.

For instance, the San Andreas fault, in its continual minor adjustments, should release a certain amount of energy that is converted into heat by friction and released into the environment. For twenty years scientists have been measuring the heat actually released by the fault and find that it is always somewhere between 10 and 20 percent of what they expect.

Of course, scientists have measured the heat released at or near the Earth's surface, and it may be that most of the heat is released a mile or two below the surface, for the big crack is a deep one. For that reason, toward the end of 1986, scientists began drilling a hole, over three miles deep, a little over two miles from the fault, in a region northeast of Los Angeles.

It may be found that the heat measured by the instruments at its bottom is fully as high as scientists think it ought to be. The question of why the heat flow is so low near the surface would then arise. Or they might find that the heat measured at the bottom is just as low as that which is measured near the surface. In that case, the problem would be why so little heat is developed and why, with so little energy, the San Andreas should be able to create monster earthquakes. Either way, the chances are that we will learn more about earthquakes and be better equipped to predict their occurrences.

The Central Heat

At what rate does the Earth grow hotter as we go deeper, and what is the temperature at the center of the Earth? These questions are important because their answers could provide clues to how the Earth formed and how radioactive materials are distributed in it. We might also be better able to estimate the internal temperatures of other worlds of the solar system and learn more about them.

We know that the Earth grows hotter as we dig deeper. We can tell this from mines and from the existence of hot springs and volcanoes. There must also be an energy source large enough to power earthquakes.

Unfortunately, legitimate estimates of the Earth's central temperatures have varied from 4,000 to 6,000 degrees Celsius (C) or 7,200 to 10,800 degrees Fahrenheit (F) and until recently there seemed no way of coming to a firm decision.

We do know some other characteristics of the Earth's interior more certainly, though. For years, scientists have been studying the waves of vibrations earthquakes create in the Earth. These waves travel in curved paths. From studying the paths of these waves we can determine the density increase of the Earth at various depths.

The Earth is rock as far as we can dig downward, but rock would not increase in density with depth quickly enough. The only materials that are markedly denser than rock are metals, and the most common metal is iron. Geologists are convinced, therefore, that the Earth has an iron "core" surrounded by a rocky "mantle."

Certain earthquake waves are known to be able to travel through solid matter but not through liquids. These waves penetrate the mantle but cannot enter the core. For this reason, geologists have decided that the mantle may become somewhat soft as the temperature increases with depth but that it remains solid. The iron core, however, is liquid.

This is not surprising. Rock melts at about 2,000 C (3,600 F) under ordinary conditions, whereas iron melts at only 1,500 C (2,700 F). A temperature that is insufficient to melt the mantle will be sufficient to melt the core.

However, this alone doesn't tell us just how hot the temperature is at the mantle-core boundary. The melting points of both rock and iron rise with pressure, and the pressure increases steadily as we go deeper. (When deep rock is forced up through volcanoes, the melting point becomes lower as the pressure drops and the volcano spews liquid rock called *lava.*)

As we penetrate more deeply into the core, the pressure continues to increase and the melting point of iron continues to rise. In fact, the melting point of iron seems to rise faster than the temperature does, so that within seventy-five miles of the Earth's very center, the iron core turns into a solid "inner core." The pressure has raised the melting point of iron so high that the temperature, even though continuing to rise, is no longer high enough to melt the inner core.

If we knew how the melting point of rock and iron increases with pressure, we would know the precise temperature required to melt iron but not rock at the boundary between the mantle and the core. We would also know the temperature at the boundary between the outer core and the inner core, for that would be the temperature of the melting point of iron at that pressure. Until recently, however, the melting point of rock and iron could only be determined at pressures far less than those deep in the Earth, making estimates difficult.

In early 1987, new techniques, in which very high pressures and temperatures can be built up for brief intervals and measured, have given us the melting point at pressures ten or twelve times as high as had previously been possible. Thus, iron melts at 4,500 C (5,700 F) at the pressure between the mantle and the outer core. It melts at 7,300 C (13,140 F) at the pressures between the outer core and the inner core.

Of course, scientists don't think the iron in the core is pure. There are other elements present, notably sulfur, and they might lower the melting point by as much as 1,000 C (1,800 F). They estimate, therefore, that the temperature is 3,500 C (6,200 F) at the outer edge of the core, 6,300 C (11,340 F) at the outer edge of the inner core, and 6,600 C (11,880 F) at the very center of the Earth.

This is hotter than had been thought. The center of the Earth is, it turns out, 1,000 C (1,800 F) hotter than the surface of the Sun.

The First Cell

Scientists are now embroiled in a discussion about what the first living cell that evolved might have been like. This is not an easy thing to decide, considering that the first living cell may have come into existence 3.5 billion years ago, and we don't have a time machine that will let us go back and look.

However, we can argue it out.

To begin with, all plants and animals are made up of cells and every one of those cells, whether in a human being, an earthworm, or a dandelion, has certain characteristics. For one thing, inside each cell is a small, more or less round object set off from the rest of the cell and containing within it the chromosomes and other materials necessary for cell reproduction. This round object is called the *cell nucleus*. All cells that possess this are called *eukaryotes* from Greek words meaning "good nucleus."

The cells in your body are eukaryotes. So are the cells in other plants and animals, and even in one-celled organisms such as amoebas. It is, however, unlikely that the first cell was a eukaryote, because eukaryotes are very complicated cells. They must have had simpler predecessors.

Even today there are simpler cells that do *not* have cell nuclei. These simple cells are very small, and the materials necessary for cell reproduction are distributed all through them. You might argue that either the cell has no nucleus or it is all nucleus. In any case such small cells without distinct nuclei are called *prokaryotes*, meaning "preceding the nucleus," because they must have come first and the eukaryotes must have evolved from them.

Bacteria are examples of prokaryotes. The best known of these fall into two groups. There are the ordinary bacteria that cannot make their own food and must live on organic materials. There are also bacteria that have chlorophyll and can make their own food. These latter are sometimes called *cyanobacteria* from a Greek word meaning "blue," because the chlorophyll gives them a bluish green tinge.

Bacteria and cyanobacteria are lumped together as *eubacteria* (that is, "good bacteria"). The eubacteria either make food the way ordinary plants do or live on organic material the way ordinary animals do, so they seem like natural organisms.

There are, however, three groups of prokaryotes that get their energy in very odd fashion and may have existed before the eubacteria. They are lumped together as *archaeobacteria* (from Greek words meaning "ancient bacteria").

The three groups are (1) *halobacteria* ("salt bacteria"), which flourish in areas with very high salt concentrations that would kill other cells of any kind and make use of sunlight as their energy source; (2) *methanogens* ("methane producers"), which live in hot springs where oxygen is absent and convert carbon dioxide into methane; and (3) *eocytes* ("dawn cells"), which live in hot springs that are rich in sulfur and produce chemical changes in sulfur compounds.

The questions are, which of these types of archaeobacteria came first and how did it develop into the others?

One way of answering such a question is to consider that all cells, whether eukaryotic, eubacterial, or archaeobacterial, contain nucleic acids. The nucleic acids are made up of chains of nucleotides, and it is possible to identify which nucleotides occur where in the chain. Very closely related species have nucleic acids with very similar nucleotide chains. In fact, it is the slow change in the nature of the nucleotide chain that produces evolution.

Scientists can make estimates of how often changes take place, and, by studying the differences in the chains, they can judge how closely related two species may be and how long ago there may have been a common ancestor. This is a difficult technique, of course.

James A. Lake at the University of California Los Angeles announced in early 1988 the results of a new computerized program that analyzed the nucleotide chains in the *ribosomes* (a cell particle essential in the production of proteins) of various types of cells.

He believes that the results show that the oldest cells

are the eocytes, and that 3.5 billion years ago, the first cells that formed were to be found in boiling hot springs full of sulfur compounds.

What's more, his results show that the descendants of these eocytes split into two branches. From one branch the other prokaryotes—the methanogens, the halobacteria, and the eubacteria—descended. From the second branch, the eukaryotes descended. In other words, we are the direct descendants of the eocytes, and the prokaryotes are our distant cousins.

Naturally, the arguments over this are going to be hot and heavy.

The Conquest of the Land

For decades, scientists believed that land life on Earth began about 400 million years ago, but a recent discovery indicates that the first land creatures, burrowers that were probably ancestors of the modern millipede, may have appeared 50 million years before that.

Earth has existed for about 4.6 billion years, and for nine-tenths of its lifetime, the land was sterile and without life.

This doesn't mean there was no life on Earth. Simple forms of life, very like the tiny bacterial cells that thrive today, existed within a billion years of Earth's formation, but they existed in the sea. For 3 billion years after that, life continued to exist only in the waters of the Earth, in rivers, pools, lakes, and oceans. The land remained untouched.

This is not really surprising, for compared with the ocean and fresh water of Earth, the dry land is an environment hostile to life.

In the sea, the temperatures are equable and vary only slightly day and night, summer and winter. On land, temperatures vary greatly, growing far warmer than the sea at some times, and far colder other times.

Water, absolutely essential to all forms of life, is always present in the sea, and sea life is in no danger of drying out. On land, water is not easily available and living things are in constant danger of desiccation. (Even human beings occasionally die of thirst.)

The buoyancy of water cancels out much of the effect of gravity so that fish can swim about in three dimensions easily. Nor does it matter how large animals become. Hundred-ton whales maneuver without trouble. On Earth, there is no buoyancy and life feels the full effect of gravity. Some small forms have developed wings and can fly about in the air (at the cost of great energy expenditure), but most land-based life can only move about the two-dimensional surface. If land animals are to move quickly, they must develop strong legs for support. Even so, large land animals are, on the whole, smaller than large sea animals.

Finally, the uppermost layers of the sea filter out harmful radiation. On land, the Sun's direct rays contain some harmful ultraviolet light that penetrates the ozone layer.

It took a long time for some forms of sea life to develop characteristics that make it possible to survive on land. Certain fish, with fleshy fins, could stump their way across bits of land to travel from one pool, in which the water was turning brackish, to another, larger one. They had primitive lungs into which they could gulp air. Slowly legs evolved, and these fish became the first amphibians. (The descendants of such amphibians today are the frogs and toads.)

This happened about 350 million years ago, and vertebrates descended from those early amphibians (including, eventually, human beings) have been living on land ever since. Amphibians had bony skeletons, and thanks to the strength this gave their structure, they could be large—the first large animals to appear on land. Some were as large as modern crocodiles.

Amphibians left fossil remains that scientists can study, but they could not have been the first animals to conquer the land. Before them came smaller animals without bones: spiders, snails, insects, and so on. It is much harder to find traces of them.

And before animals could conquer the land, there had to be food there for them to eat. Simple plants must therefore have made it to land before animals did. Until quite recently, it was thought that plant life first reached land about 400 million years ago.

In early 1987, however, two geologists from the University of Oregon unearthed evidence that showed simple land life to be older than had been thought. They dug up rocky layers in central Pennsylvania, which, as indicated by certain subtle properties, had been soil a long time ago—perhaps as long as 450 million years ago.

In that soil are burrows that do not seem to be a natural part of the soil. Increases in density toward the top and an encrustation of the walls of the burrows with certain chemicals make it appear that they were formed by burrowing animals.

From the nature of the burrows, one can deduce some of the characteristics of the animals that made them. They would have to have been animals with a long history on land, of a kind that burrowed underground, with a certain shape, a certain pattern of growth, and so on. Very likely, they are forms of life that have long since grown extinct, but the evidence seems to point to a relation to modern millipedes ("thousand legs," although the actual number is fewer than that).

And since millipedes couldn't exist without food, very simple mosslike plant life must have already existed on land before they arrived. This means that we may well have to push the conquest of land back some 50 million years. By the time our ancestors, the amphibians, arrived, the millipedes must have been there for 100 million years.

Even with this extension, however, there has been land life on Earth only for the last 10 percent of the planet's existence.

How Green Plants Began

Two groups of scientists are now locked in a controversy on the details of how green plants evolved.

This is a major question because green plants use the energy of sunlight to convert simple substances—carbon dioxide, water, and minerals—into the complex substances making up plant tissue. All animals (including us) depend on plant tissues, either directly or indirectly, to survive. Animals either eat plants or eat other animals that eat plants.

Furthermore, in forming their tissues from simple substances, green plants discharge oxygen. This is how the oxygen content of our atmosphere was formed and is maintained. And it is this oxygen that is breathed by all animal life (including us) and that keeps us alive.

Because food and oxygen are thus the gift of the green plants to the animal world (including us), anything that would explain how they came to be is of profound interest to us.

All green plants, and all animals, too, are made up of microscopic units called *cells*. The cells, although very tiny, have a complex structure, and are made up of still smaller structures called *organelles*.

For instance, all cells have *nuclei* that contain the materials of heredity that allow cells to multiply and preserve their characteristics; they have *mitochondria*, where food molecules are combined with oxygen to form energy; they have *ribosomes*, where each cell's special protein molecules are formed; and so on.

The cells of green plants have one organelle that animal cells don't have. Green plant cells have *chloroplasts* that possess the ability to use the energy of sunlight to form the food and oxygen that keep the animal world alive. Animal cells don't have them.

How did these complicated little cells, making up plants and animals, evolve? *When* did they first form?

Scientists who study the fossil records and delve back into the dim past of life-forms are of the opinion that the first cells of the complicated type that make up plants and animals formed about 1.4 billion years ago. By that time, though, the planet Earth had already existed for 3.2 billion years, so there was a long period during which those first cells might have evolved.

It is possible that before those first cells evolved there were smaller, more primitive cells that existed as the only forms of life over a period of perhaps 2 billion years. Such small primitive cells still exist today in the form of bacteria.

Bacteria are so small that a thousand of them can fit inside a typical plant or animal cell.

Bacterial cells don't have the rich collection of organelles that plant and animal cells have, and they come in different varieties. Some scientists think that about 1.4 billion years ago, different types of bacterial cells combined to form the more complicated cells that make up plants and animals. Nuclei originated as bacterial cells that specialized in hereditary control. Mitochondria were once bacterial cells that specialized in energy formation. Ribosomes were once bacterial cells that specialized in protein production, and so on.

In particular, chloroplasts were once bacterial cells that specialized in the use of sunlight for energy. Such organisms still exist today and are called *cyanobacteria*.

In one respect, though, cyanobacteria and chloroplasts are different. The chloroplasts of green plants contain two very similar key substances that are essential to the trapping of sunlight. These are "chlorophyll a" and "chlorophyll b." Cyanobacteria, however, contain chlorophyll a only. It may be that after chloroplast cells formed, they evolved chlorophyll b as a second component.

In 1985, however, a variety of cyanobacteria called *Prochlorothrix* was located in ponds in the Netherlands, and these did contain both chlorophyll a and chlorophyll b. It seemed possible that this variety was a descendant of the original cyanobacteria that became the chloroplasts found in the cells of green plants.

To check on this, it was necessary to study the fine molecular structure of *Prochlorothrix* and chloroplasts to see how similar they were.

Clifford W. Morden and Susan S. Golden of Texas A & M University studied a key protein that exists in both *Prochlorothrix* and chloroplasts. They found that the proteins of both had important similarities and that this trait distinguishes these two types of cells from other cyanobacteria. This makes it appear that *Prochlorothrix* and chloroplasts have a common ancestor.

However, Sean Turner and others at Indiana University studying nucleic acids in both *Prochlorothrix* and chloroplasts find differences that make it appear that the two are not closely related.

Obviously, more work is needed.

The technique used in these molecular analyses, however, may come to be used not only to solve this particular problem but to work out the evolutionary development of cells generally.

Dinosaurs Everywhere

Sciences are strongly interconnected. To make a discovery in one area of science is sure to shed light on other areas.

In November 1986, for instance, the Argentine Antarctic Institute announced the discovery of fossilized bones on James Ross Island, a bit of land just off the coast of Antarctica, where that frozen continent most closely approaches the southern tip of South America. The bones were unmistakably those of an ornithischian dinosaur.

The fossilized remnants of dinosaurs had already been located on every other continental landmass on Earth. The presence of these ancient reptiles on Antarctica as well makes the dinosaurs a truly worldwide phenomenon.

The discovery, however, is less important in connection with dinosaurs than it is in connection with Antarctica. How could a dinosaur live in Antarctic regions? Dinosaurs were not well adapted to extremes of cold. Indeed, back in 1968, fossilized remnants of ancient amphibian life were discovered in Antarctica. Amphibia (of which frogs and toads are the best-known modern examples) are even less well adapted to Antarctic weather.

Besides, it is not likely that dinosaurs evolved on every continent independently. If they evolved on one to begin with, how did they manage to cross the ocean to other continents?

The answer is that it is the continents, not the dinosaurs, that did the moving. About thirty years ago, it was discovered that the crust of the Earth was made up of large plates that fit tightly together but that very slowly moved. Some plates pulled apart, some plates crushed together, one plate might slowly dive beneath another. The study of such *plate tectonics* suddenly made sense out of almost everything in geology—volcanoes, earthquakes, island chains, ocean deeps, and so on—that had been mysterious before.

The plates carried the various continents on their backs, so to speak. As the plates moved this way and that, the continents moved with them. Every once in a while, the plates would bring all the continents together so that Earth would consist of one major landmass called Pangaea (Greek for "all-land"). Then, as the plates continued to move, they would pull the continents apart again.

Pangaea formed and broke up a number of times, in all likelihood, during the course of Earth's history of over 4 billion years. The last time Pangaea existed intact was about 225 million years ago. It had been intact for millions of years, but now it was beginning to show signs of breaking up.

By that time, however, the early dinosaurs had evolved

and had been given time to spread out over all parts of Pangaea. All of the landmass seemed to exist in the tropic and temperate zones so that the dinosaurs could live in reasonable comfort in the various parts of it.

By about 200 million years ago, Pangaea had broken into four parts. The northern portion was what are now North America, Europe, and Asia. To the south was a portion made up of what are now South America and Africa. Farther south still was what are now Antarctica and Australia, and a small piece that is now India.

As time went on, North America split away from Europe and Asia, and South America split away from Africa. (If you look at a map, you will see how neatly South America would fit into Africa if they were pushed together.) India moved northward and, about 50 million years ago, collided with Asia and formed the huge Himalayan Mountain range where the two landmasses came together and slowly crumpled. Antarctica and Australia also separated.

Each continent, as it separated from the others, carried its own load of dinosaurs with it. By 65 million years ago, when the dinosaurs all became extinct for one reason or another, the continents were well separated, and each one now carries its load of dinosaur fossils.

Antarctica had its dinosaurs, too, as well as amphibians and all the other plants and animals that flourished during the period of the dinosaurs. Its fate, however, was more tragic than that of others, for its plate carried it southward toward the pole. Little by little, over a period of 100 million years, it experienced a slow cooling. Slowly, plant life grew more sparse and still more sparse. Animal life thinned out. The weather grew snowier and the summers grew shorter and cooler, and finally the ice came.

Now Antarctica, nearly centered over the South Pole, is the icebox of the world. Nine-tenths of all the ice on Earth is found in the Antarctic ice cap. That ice, several miles thick, effectively covers the rich supply of fossils we would find if the soil of Antarctica were bare.

So the discovery of the dinosaur fossil in Antarctica is another strong piece of evidence in favor of the slow and inexorable geological movements of the Earth's crust.

Squashed Sand

For nine years now, scientists have been arguing over a new explanation for the disappearance of the dinosaurs 65 million years ago. But that matter may finally have been settled at last.

In 1980, it was reported that in a 65-million-year-old thin layer of sediments there was an unusual concentration of the rare metal iridium. Some suggested that it could have come from a collision or the impact with the Earth of a sizable asteroid or comet. The impact would have punctured the crust, set volcanoes to exploding, caused huge fires and tidal waves, and sent so much dust into the stratosphere as to cut off the sunlight for a long time. This would have brought about the death of much of Earth's life, including all the dinosaurs.

There's no question that 65 million years ago there was a "great dying" and that a catastrophe occurred, but not all scientists were ready to accept that it was the result of a large impact. In 1987, for instance, it was pointed out that if the Earth suddenly underwent a period of explosive volcanism, with many volcanoes erupting more or less simultaneously, this would be enough to produce a catastrophe of the size that could have caused the mass extinctions.

So things have boiled down to competing theories of "impact versus volcanism."

The question isn't just academic, because we may face one or the other catastrophe again someday (although, in the case of an object's striking the Earth, we may someday know how to prevent the impact). We need to know as much as we can about the effects of these events, so we can try to plan some kind of emergency steps to take should we face them in the future.

So, scientists have been busy trying to find evidence to support two theories.

In 1961, a Soviet scientist named S. M. Stishov found that if silicon dioxide (very pure sand) were put under huge pressure, its atoms were forced closer together and it became very dense. A cubic inch of this squashed sand weighed considerably more than a cubic inch of ordinary sand. This squashed sand has been called *stishovite* ever since.

Stishovite isn't really stable. The atoms are too close together, and they tend to move apart and become ordinary sand again. However, they are held together so tightly that this change takes place extremely slowly, so that stishovite can remain as it is for millions of years.

The same thing happens with diamonds. The carbon atoms in diamonds are pressed unusually close together, and there is a tendency for them to spread out and become ordinary black carbon, but that, too, takes millions of years under ordinary conditions.

You can hasten the change, however, if you raise the temperature enough. This adds energy to the atoms and allows them to pull away from their neighbors and resume their usual configuration. Thus, if you heat stishovite at 850 C (1560 F) for thirty minutes, it will turn into ordinary sand. (You can also make black carbon out of a diamond by heating it in the absence of air, but who would want to?)

Stishovite was made in the laboratory. Does it also occur in nature? Yes, but only under conditions where great pressure has been placed on the soil.

For instance, stishovite has been found at places where there is evidence that a sizable meteorite once slammed into the ground. The great pressure of the impact formed the stishovite. Stishovite has also been found in places where experimental nuclear explosions have taken place. The huge pressures of an expanding fireball formed it.

It seems certain that stishovite must also occur deep under the Earth's crust where the pressures are extremely high. In this case, it might be brought to the surface by volcanic eruptions. However, those eruptions are extremely hot and the rock is liquefied. Any stishovite that emerged from a volcano would be converted into ordinary silicon dioxide. And, as a matter of fact, no stishovite has ever been detected at sites of volcanic activity.

You can say, then, that the presence of stishovite indicates that an impact must have taken place and that volcanic action must not have taken place.

Well, John F. McHone and several co-workers at Arizona State University studied rocky layers in Raton, New Mexico, layers that were 65 million years old and therefore date back to the time the dinosaurs disappeared.

They made use of modern techniques of determining atomic arrangements in solid materials—nuclear magnetic resonance, as well as X-ray diffraction—and on March 1, 1989, they reported they had definitely detected the kind of atomic arrangement found in stishovite.

This makes it appear that there was a vast impact 65 million years ago that formed tons of stishovite, which was kicked up into the stratosphere before settling down to

Earth. It wasn't volcanic action that killed the dinosaurs, then; it had to be the impact.

Death of the Dinosaurs: A New Clue

A decade ago the theory was advanced that the dinosaurs (and certain other living species) were killed off, 65 million years ago, by the collision of a rather large meteor or comet with the Earth. Other scientists claim that vast volcanic reactions or other climatic abnormalities killed the dinosaurs. Nevertheless, the impact supporters have been winning, and now a new piece of evidence that may clinch the matter has arisen.

In the sediments that were laid down 65 million years ago, Jeffrey L. Bada of the Scripps Institution of Oceanography in La Jolla, California, has discovered amino acids.

Amino acids are the building blocks of proteins. Every protein molecule is made up of one or more strings of anywhere from a dozen to several hundred amino acids. In general, amino acids on Earth are produced only by living tissue.

In that case, there should be nothing particularly unusual about finding amino acids in materials that were deposited 65 million years ago. After all, there was plenty of life then and all the forms of life were forming amino acids. Why shouldn't some be found?

Well, for one thing, there are uncounted numbers of

amino acids that are theoretically possible, but the proteins formed by living organisms make use of only twenty different kinds of amino acids. What's more, all forms of life, whether viruses or oak trees or starfish or snakes or human beings, form and use the same twenty amino acids, with very rare exceptions.

Nobody knows why it is these twenty that living organisms use, and what is wrong with all the others they don't use.

The amino acids reported in the old rocks by Bada in June 1989 are, however, of two types, isovaline and alpha-aminoisobutyric acid. These are not found in proteins and, as far as we know, are not formed by living things generally. A rare form of fungus does form some isovaline, but this is very exceptional.

Is there anywhere else that amino acids are found? Well, yes. There are certain meteorites called *carbonaceous chondrites* that contain small quantities of water and carbon compounds. Among the carbon compounds are some amino acids. And, as a matter of fact, among the amino acids found in occasional meteorites are isovaline and alpha-amino-isobutyric acid. It is possible, then, that the amino acids are the result of a vast impact of a meteor or a comet that contained these amino acids and scattered them over the face of the Earth.

Can we be sure? After all, that rare fungus does form isovaline. Perhaps 65 million years ago, some animals that passed into extinction happened to make copious quantities of these amino acids that are rare now but weren't rare then.

No, we can be quite positive that this didn't happen. Amino acids, like many other substances important to life, have asymmetric molecules and can exist in either of two forms, a left form or a right form (like gloves and shoes). It so happens that the enzymes of living things make amino acids that are all left forms. Left forms fit together easily to make useful chains for the formation of protein molecules. Left forms and right forms scrambled together wouldn't work. Of course, a chain made of all right forms would work, too, but when life first began 3.5 billion years ago, the left forms were initially used through some random process, and amino acids have been left-form ever since. Even the rare fungus that forms isovaline contains only left-form isovaline.

If amino acids are formed by artificial or random processes, however, as when they are formed in the chemists' test tubes by ordinary chemical reactions, both left forms and right forms appear in equal quantities. Neither one has the advantage. The amino acids that are found in meteorites are present as left forms and right forms in equal quantities, and this tells us they were made by chemical reactions that did not involve the enzymes of living organisms.

The amino acids found in the 65-million-year-old sediments are also found in left forms and right forms in equal quantities; this is a strong sign that they were formed not by living things on Earth's surface but by nonlife processes in a meteor or comet.

Of course, there are some questions about this discovery. How is it that the amino acids were not destroyed by the heat of impact? There is no easy answer. The amino acids are not very sturdy molecules and normally cannot withstand such heat. Perhaps they existed on the inside of chunks of the impacting object and were protected from the heat.

More puzzling is the fact that these out-of-this-world amino acids don't occur exactly at the line of sediment that marks the time of 65 million years in the past. Instead, they are found a distance above or below the line. Perhaps they were originally in the right sediment layer, but, in all those millions of years, they have managed to drift through the rocks higher or lower. This doesn't sound convincing, but

Bada is investigating rocks in other areas, and perhaps additional data will supply explanations.

Fossil Fact or Fiction

Could it be that the most important fossil ever found is a fake? A few scientists claim it is and have created quite a stir.

The fossil in question was discovered in 1861; it is estimated to be about 140 million years old. It is a clear impression, in a rock, of an animal about three feet long that looks very much like a lizard. It has a head possessing teeth and no beak, a long neck, and a long tail, together with a flat breastbone; all very lizardlike.

From all this, why not deduce that the animal is an extremely ancient reptile, ancestral to the lizards of today?

This might have been the case, were it not for one all-important difference. The so-called lizard had feathers. The imprint of those feathers is present and is unmistakable. The feathers are in a double row down the length of the tail and are also present along the forelimbs.

In the world today, every known bird has feathers, and all living things that are not birds do not have them. Therefore, this fossil is thought of as the remains of a very ancient and primitive bird. It is called *archaeopteryx* (ahr-kee-OP-tuh-riks), from Greek words meaning "ancient wing."

The archaeopteryx is the best-known example of a fossil

of a life-form that seems to fall exactly between two major groups of animals as recognized today. It is half reptile and half bird and therefore a perfect example of a reptile in the process of evolving *into* a bird.

It is such a primitive bird that, at best, it may only have been able to glide. Anything more than very weak flight would not seem possible for archaeopteryx. Naturally, one might ask, what was the point of feathers if, when they first developed, they did not make flight possible. Surely, it makes no sense to suppose that useless primitive feathers would evolve simply because they might someday be useful.

The answer evolutionists give is that even if they only helped the bird glide, this in itself would be useful, and that the situation would slowly improve until full flight was possible. Indeed, feathers might not have evolved for the purpose of flight to begin with but as a sort of "net" with which to trap insects. Their use in flight and as insulation would develop as secondary characteristics later.

The English astronomer Fred Hoyle, along with two associates, claimed in 1985, however, that archaeopteryx was merely a primitive lizard and that once the fossil had been discovered, a layer of cement had been placed over it and into that cement had been pressed modern feathers intended to leave the impression of a lizard-bird.

Presumably this had been done by someone merely intent on perpetrating an amusing hoax on scientists. (Such hoaxes have indeed occurred both before and since.) Or else some enthusiastic evolutionist was eager to produce evidence in favor of evolution and didn't mind faking some for what he considered a good cause.

However, even if archaeopteryx was faked, this would not, of course, invalidate biological evolution. The truth of evolution rests not on any one fossil but on vast numbers of them and on much more besides. Even if no fossils existed, there is enough physical, physiological, biochemical, and anatomical evidence to convince scientists that evolution is indeed a fact.

Nevertheless, it can't be denied that fossils offer the most vivid testimony about evolution and that archaeopteryx is the most splendid single piece of this type of evidence.

Scientists, generally, have reacted to Hoyle's claim with anger and contempt, and Hoyle has, so far, made no converts among them. (Hoyle has also originated several other unpopular theories, such as that the universe comes into being through continuous creation rather than through the big bang and that simple forms of life actually form in cosmic clouds and in comets. Many, therefore, dismiss him as a sort of maverick who need not be taken seriously.)

The Natural History Museum in London, which possesses the archaeopteryx fossil, is convinced of its authenticity and points out the existence of tiny, even microscopic correspondences that make it seem certain that bones and feathers both made their impression on the rock at the same time. Hoyle now wants to take a pinhead of rock from the fossil and subject it to tests, but the museum will not allow it and insists that Hoyle's intended tests would prove nothing one way or the other.

I myself am not impressed by Hoyle's claims. For one thing, I doubt that a nineteenth-century hoaxer could have made the feathered impressions so well that they would fool modern paleontologists (even the famous Piltdown hoax fooled paleontologists only temporarily). Much more important, at least two other fossils of archaeopteryx have been found and they, too, have feathers arranged as in the first fossil.

Three identical hoaxes? That's harder to believe than a reptile with feathers.

More Evidence of Feathered Fliers

The most valuable fossils are those that represent intermediate forms between two well-established groups of organisms. Such fossils tend to show the course of evolution. A limestone outcrop in Cuenca, Spain, yielded what seems to be another example early in 1988. It's of an ancient bird that may be 125 million years old.

To many people one of the chief difficulties with the evolutionary notion is the question of how a complex creature can possibly evolve. Birds, for instance, have feathers; beaks; special muscles to move the wings; light, hollow bones; and many other characteristics, all of which are essential to flying and to being a bird.

How could all this develop in such a way as to produce a bird that is a working organism? Can we expect a bird to begin by developing a rudimentary wing that isn't capable of allowing it to fly? Why should such a "part wing" be developed? And if we try to imagine a bird with all the equipment for flying developing out of a lizard that can't fly, how can all that development come about all at once?

The answer would appear to be that developments are indeed made piecemeal, but the value of each development to begin with is not necessarily the same as when it is fully developed.

Consider, for instance, the "archaeopteryx," which is the first living organism we know of that we would label a bird and which appeared about 150 million years ago. The only reason we call it a bird is that it had feathers, which, nowadays, only birds have.

Aside from the feathers, though, it's a lizard. It has a lizard head with teeth in its jaw, a long tail, and so on. The feathers line the forelimbs and the tail, but it is very doubtful that the archaeopteryx could fly in the modern sense. Flying birds today all have a keel on the breastbone to which powerful flying muscles are attached, but archaeopteryx had only a small keel.

In that case, why should archaeopteryx have developed feathers? One possibility is that the feathers were a trapping device for insects. The archaeopteryx might run on its hind legs (as some lizards do today) and hold out its forelegs to catch insects. The feathers would widen the effective reach of the forelegs and entangle the insects.

The feathers, however, would also act as a parachute. If an archaeopteryx leaped, it would stay in the air a little longer because the feathers would supply more surface. If it climbed a tree and jumped, the feathers would enable it to flutter a longer distance. This would be very useful, for the higher and longer its leap, the greater its chance of escaping a predator trying to make a meal of it.

It might well turn out that this ability to jump longer and farther was so useful that any random change that improved the ability would increase the chances of the archaeopteryx's survival and enable it to produce more young that would inherit the characteristic.

Little by little, flight would improve, as other characteristics also developed: a slightly better keel to which stronger muscles could be attached, somewhat lighter bones, a more compact body, a shorter tail, and so on.

This view is now helped by the discovery of bony remains of another feathered creature in Spain. It is more recent than the archaeopteryx by perhaps 25 million years, so there was plenty of time for it to develop additional bird-like features.

This new fossilized remnant is of an organism smaller than archaeopteryx. Whereas archaeopteryx was the size of a crow, the new fossil was the size of a robin. (The smaller an organism, the easier it is for it to fly.)

The new fossil has not entirely divorced itself from its lizard forebears. Its hind limbs and its pelvic bone are quite primitive and closer to those of the lizard than to those of modern birds.

However, the fossil has a shoulder bone called the *coracoid.* In modern birds such a bone helps convert the pull of a muscle into a powerful stroke of the wing. The mere presence of this coracoid is good evidence that the fossil is of a bird that is capable of flying.

What's more, at the end of its vertebral column there is a bone called a *pygostyle,* which modern birds have at the base of their tail. This means that the fossil had a bird tail instead of a lizard tail. A bird tail has feathers that act as a brake on flight when the bird is landing—again evidence that the fossil could fly.

Unfortunately, no skull was found, so we can't tell how that might resemble modern bird skulls and what kind of beak, if any, it might have had. However, further searches may well turn up similar fossils that can answer additional questions. For now, though, we have uncovered the first bird we know of that is capable of true flight and, through it, have learned more about the development of birds.

The Biggest Flyer

About 65 million years ago, the pterosaurs, which included the largest animals that ever flew, died out quite suddenly. Their disappearance has left us with some challenging questions, including one of the most puzzling: how did these winged reptiles, some of whose dimensions approximated those of a fairly large airplane, manage to fly? Scientists remain perplexed, but several fascinating theories have emerged.

The flying reptiles called *pterosaurs* (Greek for "wing lizards") evolved as early as 200 million years ago. Though some pterosaurs were no larger than sparrows, others were the largest flying animals that ever existed. About 70 million years ago, the *pteranodon* (Greek for "wing toothless") had a wingspread of up to twenty-seven feet, almost three times that of an albatross. To be sure, it was almost all wing and may not have weighed more than forty pounds.

In 1971, however, remains of a pterosaur whose wingspan may have been as wide as fifty feet were located in Texas. It surely would have been heavier than any other flying animal that ever lived. By studying those remains and others, including fossilized hipbones recently discovered in Europe, scientists are trying to work out the puzzle.

The only clues we have about how these massive creature flew, other than the pterosaur's fossilized remains, come from examining the three other groups of flying animals that still exist.

Flying is a difficult task, and it takes concentrated energy to beat wings against the air in such a way as to rise upward and remain supported on that thin medium. Today, the only flying species that, like reptiles, are cold-blooded are the insects. Because they are cold-blooded, they produce energy at relatively low levels. They get away with flying because they are small, so small that Earth's gravitational pull on them is weak, and even thin air is sufficiently buoyant to cancel out part of that pull. The largest insect is the Goliath beetle, which weighs not quite four ounces.

Two other groups, birds and bats, are warm-blooded and can, therefore, pump far larger concentrations of energy into the task of flying. Their warm-bloodedness must be insulated, since not much of the energy produced with such difficulty can afford to be lost as radiant heat. Birds, therefore, have feathers, a particularly efficient method of cutting down on heat loss. Bats have a slightly less efficient covering of hair.

Bats and birds, thanks to their great energy production, can fly even though they are considerably larger than insects, but even so, they aren't as large as nonflying animals.

The largest bat is a fruit eater from Indonesia. It can be up to sixteen inches long and has a wingspread of nearly six feet. Its body is mostly wing membrane, however, and its total weight doesn't quite reach two pounds: eight times the weight of the largest insect.

The heaviest bird capable of flight, the kori bustard of eastern and southern Africa, may weigh as much as forty pounds, twenty times the weight of the largest bat. At this weight, however, it can just barely fly. Some albatrosses, which are not quite so heavy, have the widest wingspread: up to ten feet.

The pterosaurs had membranous wings like the bats', but whereas, in bats, the membrane stretches over all the fingers but the thumb, the pterosaurs' membrane was attached to a vastly overgrown fourth finger. The first three fingers remained as small clawed digits outside the wing.

So, how *did* they fly?

Today, all reptiles are cold-blooded and tend to be sluggish in comparison with birds and mammals. Naturally, then, it was taken for granted, at first, that pterosaurs were cold-blooded, too, and that they therefore couldn't dispose of enough energy to fly efficiently. Scenarios were worked out in which pterosaurs painfully climbed to a clifftop and then glided downward, catching their prey.

This would have been terribly difficult for them, however, and there has been an increasing tendency to view them as true wing-beating fliers. Since this takes enormous energy, more and more scientists have begun to assume that they must have been warm-blooded, meaning that they were covered with hair instead of feathers. (If they had feathers we would probably have found impressions of them on an occasional pterosaur fossil, but we haven't.)

To appreciate the difficulty flight must have posed for these creatures, consider the May 1986 demonstration at Andrews Air Force Base near Washington, D.C., of a pterosaur like the one discovered in Texas. The giant forty-four-pound model, which had a wingspan of eighteen feet, was constructed for the Smithsonian Institution at a cost of $700,000. It was airborne for only one minute before it fell apart and crashed to the ground before a crowd of Armed Forces Day spectators.

But even if pterosaurs somehow did manage to fly, how did they walk? Two sets of pterosaur hipbones were located in West Germany, and they were not badly crushed. From these it can be deduced that the pterosaur's thighbones spread outward. In that case, it is likely that the pterosaurs waddled on the ground and were clumsy walkers. We might, therefore, further deduce that, when not flying, they hung from trees or cliffs.

They were, in other words, like gigantic bats in many ways, but their bone structure was that of reptiles, and in all likelihood, they laid eggs rather than giving birth to living young, as bats do.

Past Monsters

In November 1987 it was announced that a new monster of the past had been discovered. A skull and other bones of a seabird, distantly related to modern pelicans, were discovered in some 30-million-year-old rocks during excavations for an airport in Charleston, South Carolina. The ancient animal was dubbed "pseudodontron."

Compare this monster with the largest seabird now alive, the wandering albatross. Whereas the albatross has a wingspread of eleven feet, the ancient pelican had one of perhaps as much as nineteen feet. Even this pales, however, when we consider that a now-extinct vulture, the largest flying land bird known to have existed, had a wingspan of up to twenty-five feet. The largest known pterosaurs (flying reptiles that lived 65 million or more years ago) had wingspreads of up to forty feet.

There is a question of how these creatures could possibly fly. The heaviest flying bird alive today, the kori bustard, weighs forty pounds or so and flies with difficulty. Albatrosses, for all their wingspread, are lighter, weighing perhaps no more than twenty-two pounds, and they have difficulty taking off. They spend most of their time in the air soaring, rather than actively flying, taking advantage of rising air currents rather than their wing muscles.

The extinct pelican, however, may have weighed as much as ninety pounds, and from its bone structure one suspects that it could only move its wings up and down but couldn't flap them to get forward thrust. Undoubtedly, the pelican could only soar, but in that case how did it get high enough to soar in the first place? How did it take off? It's a puzzle. That goes for the extinct vulture and for the large pterosaurs, too.

Another puzzle is why animals that existed in the past were so much larger than any now living. The largest living primate is the lowland gorilla, which is as tall as a man and may weigh up to four hundred pounds. A few million years ago, however, there lived a still larger species, *Gigantopithecus* ("giant ape"), that was up to nine feet tall and might have weighed close to one thousand pounds.

The largest land mammal now alive is the African elephant, which stands nearly eleven feet tall at the shoulder and can weigh up to six tons. However, about 20 to 40 million years ago, there lived a giant rhinoceros (without nose horns), the *Baluchitherium* ("beast from Baluchistan"), which stood nearly eighteen feet tall at the shoulder, the height of the tallest giraffe. It was up to twenty-eight feet long from head to tail and weighed at least twenty tons.

And about 150 million years ago there lived a giant dinosaur called *Brachiosaurus* ("arm lizard"), which was the largest land animal of any kind. It was twenty-one feet tall at the shoulder and had a long neck that could lift its head as high as forty feet in the air, tall enough to look into the fourth-story window of a modern building. It might have weighed as much as eighty tons, thirteen times as much as the largest elephant alive today.

Or consider birds. The largest birds alive today are far

too heavy to fly, but they can do well anyway. The ostrich holds the record for living birds. Its head, atop its long neck, can be eight feet off the ground. It can weigh up to 280 pounds, and it can run at speeds of up to forty miles per hour.

Yet only a few centuries ago, there lived giant ostrichlike moas in New Zealand that held their heads at heights of up to thirteen feet and weighed up to five hundred pounds. This is a height record, but not a weight record. Up to the 1600s, the "elephant bird," or *Aepyornis*, was still alive in Madagascar. It stood only ten feet high, but it weighed nearly one thousand pounds. It had the largest egg known, with a capacity of two and one-third gallons, seven times that of an ostrich egg.

Or consider insects. There are beetles up to seven inches long, and some that weigh as much as three and a half ounces. But about 300 million years ago, there were dragonflies with bodies over a foot long and a wingspread of up to two and a quarter feet.

Is life degenerating, then? I don't think so. I think that in the long run, it just turns out that small and nimble works better than large and clumping. Besides, we have our own wonders today.

After all, the largest animal of all time is alive today. It is the blue whale. It can be up to ninety feet long and weigh up to 130 tons, twice the length and size of the largest dinosaur that ever lived.

Nor is it likely that there were ever any taller trees than the redwoods of today (along with some other species), which reach heights of four hundred feet or so. And the most massive trees (and therefore the most massive of any form of life) are also alive today. These are Sequoia trees, the largest of which may weigh up to sixty-seven hundred tons (fifty times as heavy as the largest whale).

Finally, the most intelligent species of all time, the only

one capable of developing philosophy, science, technology, art, literature; the only one to make fire, use electricity, and stand on the Moon, is alive today. It is *Homo sapiens*, and, as geology counts time, we only came into existence yesterday.

The Most Successful Life-Form

The American Museum of Natural History in New York bought a mineral collection from Columbia University in 1980. The mineral collection included pieces of amber, and in late 1987, museum curator David Grimaldi was looking over the bits of amber and found himself staring at a bee that was 80 million years old.

It may not seem like much to those of us who dismiss insects as annoying pests, but the fact is, insects are the most successful forms of life on Earth. An alien from a distant planet who was examining Earth dispassionately might well report to his superiors that Earth was an insect world with an unimportant scattering of other types of life.

Consider that almost a million different species of insects are known. This is a far greater number than the total of all other species of life *combined*. In fact, about five out of every six species of life on Earth are insects.

What's more, this only includes the life-forms that are known. There are many millions of species that have not yet been discovered, named, and described—especially in the tropical forests—and of these it is confidently estimated that almost all are insects of one sort or another. There may be anywhere from 2 million to 5 million species of insects actually in existence, and it is possible that up to 97 percent of all species are insects.

Why are insects so successful? They are small and they are fecund, laying uncounted numbers of eggs. There may be as many as 4 million individual insects in one moist acre of land.

This means that insects are difficult to wipe out. Kill ninety-nine out of one hundred and those who manage to survive will lay enough eggs to restore the population in no time. In fact, although human beings have easily driven to extinction various large forms of life such as mammoths and mastodons and continue to endanger many others, it would appear that we have never managed to wipe out a single insect species. Roaches and mosquitoes flourish, for instance, though every person's hand is against them.

The vast turnover in numbers means that evolution proceeds at an enormous rate among insects, so that new species with new characteristics are being formed much more quickly than among the other animals about us. Every once in a while we slaughter insects with insecticides and kill billions of them. A relatively few, however, just happen to be naturally resistant to a particular insecticide. They survive and rapidly have millions of descendants, all resistant. In the space of a few years, the insecticide loses its value and a new one has to be found.

Paleontologists would dearly love to have the details of the evolutionary development of insects, but they are small and leave few fossils. The oldest traces of insects are those of very primitive "springtails," insects that have no wings and can do no more than spring in an attempt to get away when frightened. (Such springtails still exist today.) Springtails lived at least 370 million years ago.

About 280 million years ago, giant dragonflies had evolved with a wingspread of up to twenty-seven and one-half inches, the largest insects that ever lived.

But the evolutionary record of insects is full of gaps. And yet we have one lucky break. Occasionally insects were trapped in the sticky resin exuded by ancient (now-extinct) evergreen trees. The resin fossilized into the material we now call amber and the amber kept the insect bodies entombed for millions of years. The oldest insects thus found are 120 million years old.

The bee that was found in the amber that originated in Burlington County, New Jersey, isn't quite that old, but it is twice as old as any other bee relic ever located. Yet, even after 80 million years, it can be seen clearly and in detail.

The surprise is that even though it is 80 million years old, it is an advanced bee, not very different from those alive today. It is a stingless honeybee belonging to a family of bees that still exists in tropical regions. Presumably, New Jersey, 80 million years ago, was considerably warmer than it is now.

To have evolved to that advanced stage of development 80 million years ago, bees may have formed perhaps another 80 million years before that. This is of importance in connection with plants. It is thought that flowering plants evolved along with bees (and similar insects), for the two are connected. Bees live essentially on the nectar of flowers, and flowers reproduce because bees carry pollen from one flower to the next. Flowering plants are thought to have originated about 135 million years ago, but if bees are older than that, then so are flowers.

Paleontologists continue to look. Every insect found in amber is precious.

The Homing Turtles

Scientific investigation can't always tell you exactly what is true, but sometimes it can tell you that something that is dramatic and sounds good may not be true. That happened early in 1989 in connection with, of all things, green turtles.

Many animals routinely migrate, breeding in one place and feeding in another place that may be thousands of miles away. This means they must find their way from one place to the other without benefit of anything more than their senses and their instincts.

Green turtles, for instance, feed along the shores of Brazil, but toward the end of each year, something drives them eastward across the Atlantic Ocean on a journey that takes two months. They finally end up on the beach of Ascension Island, a small bit of land in the mid-Atlantic about twelve hundred miles east of Brazil. There they nest and breed and then return (another twelve hundred miles, another two months) to Brazil to feed until the end of another year drives them eastward again.

The little turtles that hatch out on Ascension Island and survive also swim back to Brazil, only to return at the end of the year. Actually, it is common for animals to travel long distances to return to the place where they were born, when they want to breed; it's called *natal homing*. Biologists speculate on just how these animals manage the navigation, but what really puzzles them is, why?

Why should turtles make such a long journey? What has Ascension Island got that other places don't have? Actually, some green turtles do nest in other places. There's one place off Florida and one off Venezuela, but Ascension Island is the most popular.

Back in 1974, Patrick Coleman and Archie Carr, two Florida biologists, made an interesting suggestion. Ascension Island is very close to the Mid-Atlantic Ridge, where, some 40 million years ago, Africa and South America nearly touched. In those days, green turtles would feed off Brazil and swim a couple of miles to Ascension Island and breed.

However, the Atlantic Ocean was forming, because material was oozing up from the ridge and forcing the landmasses apart (*seafloor spreading*). Every year, the feeding grounds off Brazil were an inch or so farther away from the ridge and from Ascension Island. Every year, the turtles swam an additional inch to get to the Ascension beaches. At no time did the turtles have any sense that the beach was perceptibly farther away, but after 40 million years, they were making a twelve hundred mile trip each way.

There's something so dramatic about the possibility of seafloor spreading's fooling the turtles that there was a strong tendency to believe the suggestion, but Stephen Jay Gould of Harvard was one skeptic. He said that in 40 million years there must have been times when Ascension Island had no beaches or when it may even have been temporarily beneath the waves for a few hundred years. That would have broken the spell.

Is there any way of testing the matter and seeing whether the green turtles have been visiting Ascension Island for 40 million years—or not?

Two scientists at the University of Georgia, Brian W. Bowen and John C. Avise, along with Anne B. Meylan of the Florida Institute of Marine Research, have been checking the nucleic acid molecules in the mitochondria of turtle cells.

These are inherited through generations and slowly change with the years (this is what makes evolution possible). The turtles that visit Ascension have nucleic acids that slowly change, and so do the turtles that nest off Florida and those that nest off Venezuela.

If, however, the turtles have always clung to their breeding grounds for tens of millions of years, each group would have undergone a different set of changes and the three sets of nucleic acids ought now to be far apart. Scientists nowadays have ways of telling how far apart they ought to be.

It turned out that differences did indeed exist among the three groups, but they were far smaller than would be expected of a multimillion-year separation. The difference was of a size that suggested a separation of only 40,000 years rather than 40 million.

That gives rise to two possibilities. In one case, the homing instinct isn't infallible after all. A certain small fraction of turtles get confused and end up at the wrong beaches. They breed and mix their nucleic acids with those who belong there. Even a small rate of "leakage" of this sort would wipe out most of the difference.

The other possibility is that turtles in general are more flexible than we think. Forty thousand years ago, some may have accidentally discovered the Ascension Island beaches, which, for all we know, may have just formed then and may not have existed before. These turtles colonized them and used them ever since, and other turtles used other beaches.

If scientists could somehow determine whether Ascension Island has been in uninterrupted existence for 40 million years, this would help them determine which theory of the green turtles' behavior is correct. But it seems probable that there would have been times when the island was submerged and that the turtles are indeed flexible.

And if, someday, the Ascension beaches become unusable, the turtles may shift to others. A less dramatic theory, perhaps, but more sensible.

The Oddest Mammal

The oddest mammal we know seems to be even odder than we had thought, and it was so odd to begin with that zoologists wouldn't believe it when a stuffed specimen first arrived in England in 1800. It had come from the largely unexplored continent of Australia.

The animal in question, still a viable species, is nearly two feet long and has a dense coating of hair that makes it clearly a mammal, for only mammals have hair. (It also feeds its newborn young on milk, and only mammals produce milk.) However, it has a flat, rubbery bill rather resembling that of a duck, something no other mammal has. It also has a spur on each hind ankle intended to secrete poison, something no other mammal has.

Although it is warm-blooded, this particular mammal doesn't maintain an internal temperature as steady as that of other mammals. Then, too, under its tail, it has one opening for wastes as birds and reptiles have instead of the two openings that mammals have. Furthermore, certain fine points in the structure of the skull resemble those of reptiles rather than mammals.

The oddest thing of all—something not definitely discovered till 1884—is that the creature *lays eggs* through that one opening in the rear, just as birds and reptiles do.

The common name of this mammal is "duckbill platypus." The "duckbill" is obvious and "platypus" is Greek for "flat feet." It is also called an *ornithorhynchus* from Greek words meaning "bird-beak."

It is not the only egg-laying mammal, for two closely related species of "spiny anteater" (also called *echidna*) live in Australia and New Guinea. These three species of egg-laying mammals are called *monotremes*, from Greek words meaning "one opening."

The monotremes (along with marsupials, such as the kangaroo, which bring forth living but quite undeveloped young) seem to be the last relics of primitive mammals that have evolved only partway from the reptile stage. They have survived only in Australia and New Guinea, which split off from the other continents before more advanced mammals had evolved.

Nevertheless, the monotremes have not stopped evolving. They may not have ever developed certain mammalian characteristics, such as the placenta, which allows advanced mammals to bring forth well-developed living young, but they have developed unusual and advanced features of their own. The poison spur of the platypus is an example.

Then, too, the platypus is a freshwater creature. Although it breathes air as all mammals do (and as birds and reptiles do), it spends much of its time on river bottoms searching for the small forms of water life it feeds on: shrimps and so on.

The question is how the platypus finds its food, as the rivers are often murky and sight becomes relatively useless. In fact, when it is underwater, the platypus shuts its eyes, ears, and nose, so it would seem that it could only find food by the sense of touch. However, it heads for shrimp and other items of diet before it gets close enough to touch them.

In 1986, scientists at the Australian National University of Canberra discovered that the platypus has a special sense. There are certain nerve endings along the edge of the platypus's bill that are sensitive to tiny electric fields. (It may be that the other monotremes, the echidnas, also have such electric sensors.)

If this is so, chalk up another oddity for the monotremes, for no other mammals seem to have such sensors. No reptile has them either, although some fish do.

The platypus's electric sense is more versatile and useful than those possessed by fish. Some fish respond only to electric currents that flow steadily in one direction; others respond only to electric currents that alternate directions. The platypus, however, responds to both. What's more, the platypus's electric sensors are connected to a different nerve from those of fish. This shows that the platypus did not build on the fish's ability but evolved its electric sense independently.

What is the use of such a sense? Well, all living animals have nerves and muscles and, when animals move, those nerves and muscles become active with tiny electric currents that flow along them. The platypus detects the field these currents give rise to and the direction from which they come, allowing them to "see" their prey by electricity.

In addition, the friction the flowing water of a stream makes against the river bottom creates tiny electric fields of its own, and the platypus may detect those fields as well. This would give it a notion of the unevenness of the bottom and allow it to maneuver safely without the use of the ordinary senses.

On the other hand, this electric sense may account for the fact that it is difficult to keep platypuses in captivity. The water has to be circulated with electric water pumps, and this may overstimulate the electric sense and prevent the platypus from flourishing.

Old Water

Are old salt mines safe places to store nuclear wastes? How can we guarantee that after thousands of years, groundwater won't seep into the mines, rust the containers, leach out the wastes, and spread them far and wide through the soil, contaminating all they touch?

One way of checking on this unknown future is by considering the past. Salt mines have formed because in the distant past, shallow arms of the ocean, heated by a pitiless Sun under conditions of little rainfall, gradually dried up.

As the ocean dried, there was too little water to keep the salt content dissolved. A crust of salt crystals, crystals that grew quickly in the heat of the day, formed. More and more salt formed, less and less water existed, until there was a totally dry salt flat. Then as the years passed, dust and sand blew over the salt and, eventually, the salt layer was buried deep under the gathering soil and, behold—a salt mine.

But while the salt crust was falling, an occasional rain would drop fresh water on the salt. This water might, for a little while, dissolve tiny quantities of the salt, but then the clouds would pass and the hot, unrelenting Sun would shine down and the water would evaporate.

At times there would be a race between the evaporation of the water and the growth of the salt crystals. On occasion, a crystal would form so rapidly that it would grow around a droplet of water or, for that matter, a droplet of the drying ocean. Nowadays, it is possible to find crystals in salt mines that contain tiny droplets of water that may date back to the ancient days when an arm of the ocean dried up—perhaps hundreds of millions of years ago.

Does it matter whether we have this old, old water? After all, water doesn't change with age, does it? No, it doesn't change with age, but it does change with evaporation.

Each water molecule is made up of two hydrogen atoms and an oxygen atom. Each hydrogen atom has an atomic weight of 1 and each oxygen atom weighs 16, so that the water molecules weigh 1 plus 1 plus 16 or 18. However, a very few hydrogen atoms (one out of sixty-five hundred) weigh 2. In addition, one out of every five hundred oxygen atoms weighs 18, and one out of every twenty-five hundred weighs 17. As a result there are a very few water molecules that, instead of weighing 18, may weigh 19, 20, 21 or 22, depending on how many atoms of heavy hydrogen and heavy oxygen they manage to incorporate.

Nowadays, it is possible to determine the average weights of the molecules in a batch of pure water with great precision and to a number of decimal places. Because water always contains the same small percentage of these heavy atoms, the average molecular weight remains very slightly above 18, whether the water comes from your faucet or the middle of the ocean.

When water evaporates, however, the lighter molecules evaporate just a tiny bit faster. (Being lighter, they can pull away from the bulk of the water more easily.) Consequently, if a considerable amount of water evaporates, the bit that remains is richer in the heavier molecules and the average weight of those molecules is measurably higher than it would be in fresh water.

Now consider the crystals of salt that contain water droplets. One can estimate how long a time has passed since particular salt crystals formed from the drying ocean. Geologists at Arizona State University, working with such salt, have a crystal that is 400 million years old, and the droplet of water it contains is the oldest sample of water yet known to exist. Samples of salt from one of the places in which nuclear wastes may be stored are 250 million years old.

The water contained in these salt crystals was evaporating at the time it was trapped. Therefore, it should be richer in the heavy atoms and have a distinctly higher average weight to its molecules than fresh water would have. At least this would be true if the old water had remained untouched. If fresh water from the soil had leaked into the salt mine and somehow found its way into salt crystals, then it would be relatively new water. It would not have evaporated much in the cool darkness of the mine, and it would have a low average weight to its molecules.

The Arizona State geologists have detected water in salt crystals by lighting those crystals from behind and studying them under a low-power microscope. In hundreds of cases they have extracted the tiny droplets and checked the weight of the molecules.

The analyses show that the molecules tend to be heavy, leading to the conclusion that the water in the crystals has remained untouched for hundreds of millions of years. Therefore, the salt mines may indeed be safe repositories for nuclear wastes.

Lightning and Life

Acid rain isn't always bad. There are some kinds that are not only good but play an active role in preserving land life. And recent experiments have made it seem that some acid rain is even more life-giving than we had thought.

Here's how it works:

One of the important elements that is to be found in all the key molecules of living tissue is nitrogen. Nitrogen atoms exist in the soil as part of mineral substances called "nitrates." Nitrates are, one and all, easily soluble. That means that when plants absorb water from the soil, that water contains a little nitrate. The plant uses the nitrates as raw material for the formation of nitrogen-containing compounds of importance to life: notably protein and nucleic acids.

Animals that eat plants (or other animals that have eaten plants) break down those proteins and nucleic acids into simpler building blocks, absorb these, and build them up again into their own varieties of proteins and nucleic acids.

All of land life depends on those nitrates in the soil. However, because the nitrates are soluble, rain tends to wash them away into brooks, then rivers, and finally the sea. In time, all the nitrates would be gone, and although life would continue in the sea, where the nitrates end up, the land would become an absolute desert—unless the nitrates could be replaced.

Four-fifths of Earth's atmosphere is pure nitrogen. If a

little of this nitrogen could be *fixed,* that is, combined with other elements, it could then be used by plants. But nitrogen is standoffish and combines with other atoms only with the greatest difficulty.

Yet the nitrates still exist in the soil and they *are* replaced. How? For one thing, human beings have learned to fix nitrogen in large quantities and the resulting nitrates can be used as fertilizer, but this capacity is recent; it was developed only about three-quarters of a century ago. How did life manage before then?

As it happens, there are bacteria that have the unusual capacity of forcing nitrogen in the air into combination with other atoms. They are called *nitrogen-fixing bacteria* and are found, particularly, in nodules attached to the roots of leguminous plants such as peas and beans. These bacteria are vitally important to life generally (including us).

And then, there is lightning. Whenever a lightning stroke flashes through the air, it momentarily heats the air around it to unusually high temperatures. The air quickly cools down, but before it does so, the heat forces the combination of molecules of nitrogen and oxygen in the air, forming nitrogen dioxide. This dissolves in water (it is usually raining at the time) to form nitric acid, which produces a kind of acid rain. When the nitric acid reaches the ground it is converted into nitrates, and this helps fertilize the earth and makes land life possible.

Until recently it was thought that lightning was responsible for about 10 percent of the nitrates in the soil. This figure was arrived at by studying simulated lightning flashes produced in the laboratory. But two American scientists, Edward Franzblau and Carl Popp, from the New Mexico Institute of Mining and Technology, have now turned to nature itself. Recently, they worked out a method for calculating the amount of nitrogen dioxide formed in actual lightning flashes during thunderstorms.

After studying about sixty flashes, they calculated that each flash produces about a thousand trillion trillion molecules of nitrogen dioxide. This is about one hundred pounds of the material. And, on the average, about one hundred lightning strokes strike the Earth per second. This means that lightning is producing five and a half tons of nitrogen oxide per second. Franzblau and Popp calculated that this means that lightning supplies the Earth not with 10 percent of the nitrogen oxide that life consumes but 50 percent.

That's impressive (always assuming that their observations and calculations are borne out and are found to be correct) and certainly gives us a new perspective on the lightning bolt. Dangerous as the bolts are, killing people and setting forest fires, it would seem that the good they do must far outweigh the evil.

Yet if the acid rain produced by the lightning bolts is so essential to life, why has the expression "acid rain" become so fearful? Why should it be a killer?

That is because the acid rain we fear is caused by the burning of impure coal and oil that contains both sulfur atoms and nitrogen atoms. Human-made acid rain contains sulfuric acid as well as nitric acid. Sulfuric acid is particularly harmful and is not a component of lightning acid rain.

In addition, human-made acid rain is considerably more acid than lightning acid rain. Where it falls there is a considerable oversupply of acid. It is this oversupply that withers forests and kills the fish populations in ponds and lakes.

Overkill

The wildfires that consumed much of the American West during the summer of 1988 were a disaster, but they were nothing compared with a much greater fire that may have happened 65 million years ago.

That age, when the dinosaurs disappeared, has become a familiar topic of scientific discussion in recent years. Many scientists have speculated about the possibility that a dramatic catastrophe from outer space was the cause of the dinosaurs' extinction.

In the rocky layers that were laid down about 65 million years ago, there is a surprisingly large quantity of the rare metal iridium. Iridium is very rare in the Earth's crust, but it is more common in meteorites and comets. This 65-million-year-old iridium layer has been found wherever scientists have looked, so the supposition is that a large asteroid or comet struck the Earth 65 million years ago and killed off most of the living species on Earth, including the dinosaurs.

Why were they killed all over the Earth, if the asteroid hit only in a single place? The first answer was that the asteroid kicked up a huge cloud of powdered rock, soil, and dust that spread throughout the upper atmosphere and blocked the light of the Sun for months. This would have killed most of the world's plant life and animal life, too, because animals depend on plants for food, directly or indirectly.

If that weren't enough, such a large strike was bound to make itself felt in other ways. For instance, 70 percent of the Earth's surface is covered by water, so that the asteroid is likely to have splashed into the ocean. It would, of course, penetrate right through to the ocean floor and still kick up the deadly dust that would block the sunlight. It would, however, also splash the water, forming a huge *tsunami*, or tidal wave.

A group of geologists led by Joanne Bourgeois were studying the rocky layers in eastern Texas and came across a thick layer of sandstone of over two feet. In it were fragments of seashells, wood, fish teeth, and so on. In the upper region there were signs of ripples that might have been caused by waves. And all of this was laid down about 65 million years ago.

The geologists therefore suggested, in 1988, that the asteroid or comet that struck the Earth at that time may have plunged into the Gulf of Mexico. The sandstone then would be the result of the huge wall of water that splashed onto the shores of the Gulf and was slowly sucked back, creating havoc. Such a tsunami would expend its force only over a limited portion of the Earth. But if the sandstone layer is being correctly interpreted, it at least lends force to the suggestion of a huge strike from outer space.

The asteroid, however, may have had an even worse effect. Recently, Edward Anders of Chicago University and other researchers reported on that key layer of rock laid down 65 million years ago in places as diverse as Switzerland, Denmark, and New Zealand. Wherever they looked, they found a layer of soot, anywhere from one hundred to ten thousand times as concentrated as seemed reasonable to expect.

This soot seemed most likely to have been the remains of a fire. By studying the exact nature of the soot, the quantity of carbon present, and the proportion of different atomic varieties (or *isotopes*) it contained, the geologists came to the conclusion that it was all one fire, taking place at one time. The fire was something that was global in nature.

Here is the scenario. The huge inrushing object from

outer space must have punctured the Earth's crust. In addition, in order to send up such a vast cloud of dust and walls of water, it also must have allowed the heated rock (*magma*) beneath the crust to spew upward. The magma may have risen both at the site of the primary puncture and elsewhere on Earth as the strike cracked and shoved the Earth's crustal plates.

Enormous volcanic activity would have set fires in many different places at the same time, and all may have joined in a roaring, more or less worldwide land surface conflagration. The soot is rich in organic compounds, indicating that vast quantities of microscopic life died at the time.

There is also the possibility that enough carbon dioxide was produced to bring about a warming trend through the greenhouse effect. Nitrogen oxides might have been produced, resulting in a vast period of acid rain. Carbon monoxide and various other small carbon-containing compounds that poisoned the atmosphere for a while must have been formed.

It's a dreadful picture, and I wonder whether it can be entirely correct. I can't help but see it as a kind of overkill. What with dust in the upper atmosphere, walls of water, continental fire storms, carbon dioxide, acid rain, poisons of all sorts, the catastrophic scenario is simply too much. If it all happened 65 million years ago, then, of course, that's what killed the dinosaurs. The question then becomes, how could *any* life survive? How is it that we ourselves are here today?

The Ozone Hole

The amount of ozone in the upper atmosphere has decreased to half of what it was about fifteen years ago. In 1985, it was discovered that a hole in the ozone layer forms near Antarctica in the fall. Data gathered by satellites over the past few years were quickly studied, and it appears that the ozone hole has been getting larger year by year. Eventually it may be that there will be no ozone layer at all in the upper atmosphere.

Does it matter? Yes, it does. The ozone layer prevents the Sun's dangerous ultraviolet rays from penetrating to the Earth's surface, where they could cause grief to human beings in the form of serious sunburns, skin cancers, cataracts, and other maladies. Even more distressing, they could kill soil bacteria and algae that populate the sea, a critical link in the ecological chain.

The problem with ozone was predicted, in the early 1970s, by two scientists at the University of California who pointed the finger at chlorofluorocarbons (CFCs), of which Freon is the best known example. These CFCs aren't inflammable. They aren't toxic. They are absolutely safe to use. They were easily liquefied and vaporized so that they could be used to transfer heat from one place to another. As a result, after World War II, they were increasingly used in refrigerators and air conditioners (and also in spray cans).

Eventually, all CFCs leak out of wherever they are and into the atmosphere. Several million tons of these have already leaked into the air and more CFCs escape every day.

In the atmosphere, they stay. They aren't washed out by the rain; they aren't changed by any chemicals; they just drift upward steadily into the stratosphere.

In the stratosphere, sunlight slowly breaks down the CFC molecules, liberating chlorine atoms. The chlorine atoms then destroy the ozone molecules.

When this danger was first pointed out, the United States banned the use of CFCs in spray cans, but they are still used elsewhere for the purpose. Furthermore, there are no good substitutes, so far, for refrigerators and air conditioners.

This is a matter of grave concern because the ozone layer in the upper atmosphere is opaque to ultraviolet light. Most of the ultraviolet light of the Sun is absorbed in the ozone and very little reaches the Earth's surface. As the ozone layer thins, more ultraviolet light will reach the Earth's surface. This means more skin cancer among human beings. There are estimates that, in the course of the next century, there will be 40 million cases of skin cancer among Americans alone, and 400,000 deaths. There would also be increases in cataracts and other conditions.

This will affect those with fair skin much more than those with pigmented skin. Therefore, Europeans and their descendants will suffer more than Africans, Asians, Aborigines, and American Indians. Among Europeans, blonds will suffer more than brunettes. (People like me, who burn to a red crisp after fifteen minutes in today's Sun with the ozone layer in existence, will be in serious trouble.)

But is that all it is, a greater tendency to burn that damages the skin? In that case can't we just stay indoors as much as possible and carry sunshades when we must be out-of-doors?

Unfortunately, the problem may well be worse than that.

Ozone is an active form of oxygen. Ordinary oxygen

molecules contain two oxygen atoms each. Ozone molecules contain three oxygen atoms. There can't be an ozone layer then until there is ordinary oxygen in the atmosphere, and such oxygen didn't build up to significant amounts until a billion years ago. For at least 2.5 billion years before that, life on Earth existed without oxygen in the atmosphere and without an ozone layer. Life-forms then were simple cells of bacterial size that lived in the sea below the uppermost layer where ultraviolet light could penetrate. There was no life at all on land, for land was exposed to the ultraviolet light.

A billion years ago, with oxygen finally present in the air, enough energy was available to allow life-forms to become complicated. It was, apparently, only about 400 million years ago (when life was over 3 billion years old) that there was finally enough oxygen in the air to allow an ozone layer that was thick enough to protect land life to develop. Only then did life move into the uppermost layer of the ocean and invade the land, too.

If the ozone layer now thins drastically, and if ultraviolet light pours down strongly, this might not affect higher plants and animals too much. We have hair, feathers, scales, cuticles, skin, bark, and so on, to protect us. But what about the soil bacteria and the sea algae that are still bare and defenseless?

Is the ultraviolet light likely to kill them? It might. The land in the uppermost layer of the sea might become as unlivable for them now as it was a billion years ago. And if these microorganisms go, we don't know how seriously this will affect higher organisms that depend upon them ecologically. In short, it may be that the very fabric of life will be seriously disrupted.

What do we do? We must stop using CFCs, obviously. This is comparatively easy. But we must also try to figure out a way of neutralizing the stuff that is already in the air, and that is much harder.

The Last Clean Place

Antarctica is the last clean place of any size that's left on Earth, and its northernmost edge was subjected to an oil spill on January 29, 1989. Things may get worse too, for Antarctica (believe it or not) is expected to get increasing numbers of tourists, along with ships, and garbage, and pollution in general. It may seem that this is one of the things we ought not to worry about. After all, isn't Antarctica just a huge waste of ice?

Actually, that isn't all it is. The interior, admittedly, is the most life-free spot on Earth, but the rim of the continent is home to penguins, seals, and skua gulls. And more than that, we must consider the area around it, the Antarctic Ocean.

Life on land depends on water rather than oxygen. Oxygen is freely available wherever one can breathe reasonably clean air, but water is unevenly distributed. There are places on land that get very little water and those places are deserts where life is rare. There are places where rain is copious and there we have forests that are filled with a riot of life.

For life at sea, it is just the other way around; it is oxygen that counts, not water. Water is freely available everywhere in the sea, but oxygen is not. Oxygen from the air and oxygen formed by microscopic green plants in the ocean surface dissolve in water, and it is these plants dissolved in the water on which all animal life in the sea must live.

The winds see to it that the surface of the ocean is roiled up, so that oxygen is constantly being dissolved. Water currents in the sea carry this dissolved oxygen to all parts and to all depths, even to the very bottom. Still, how much oxygen is that? Is it enough?

This depends on the temperature. The warmer the temperature of the water, the less gas of any kind it will dissolve; and this, of course, includes oxygen. This means that the warm water of the tropical seas dissolves only a little more than half as much oxygen as the icy water of the polar seas.

For human beings exposed to the cold water near the poles there is only numbness and quick death, whereas the warm water of the tropics is a paradise for swimmers, but not so for sea life in general.

The tropical waters, with their dearth of oxygen, are, compared to other parts of the ocean, deserts. The polar waters, however, are rich with life. Microscopic life swarms in incredible abundance. Small animals feed on it; larger animals feed on them, and are eaten by still larger animals, and so on.

In fact the largest animal that ever lived on Earth, past or present, the gigantic blue whale, which can weigh up to 150 tons, twice the mass of the largest dinosaur that ever lived, makes its home in the Antarctic. It lives by scooping up, in its gigantic maw, vast quantities of krill, small shrimplike animals up to two inches long.

So if the continent of Antarctica itself is largely a frozen wasteland, the waters surrounding it are the richest reservoir of life on the planet, supporting the familiar large animals of the far southern seas: the penguins, gulls, seals, dolphins, and whales. If we damage the region badly, we tear an enormous hole in the fabric of the planetary ecology. Remember that every part depends on every other part, so that damage in the one place that is richest in life is bound to result in some damage everywhere.

Many kinds of pollution are cleaned up gradually by natural processes. An oil spill, damaging though it is, evaporates to some extent, breaks down, is gradually degraded, and finally disappears. It doesn't do this quickly enough, to be sure, so that before it degrades, it causes much harm. In this Antarctic incident, several hundred tons of diesel fuel was spilled. Diesel fuel spreads out and evaporates more rapidly than fuel oil—but it is also more toxic.

Unfortunately, all chemical processes tend to slow down as temperatures drop, so that, in cold water, the evaporation and degradation of the spill are slowed. And, at cold Antarctic temperatures, the rate of evaporation and degradation may be only one one-hundredth what it is in warmer climates. It may be, then, as some pessimists calculate, that spills of this sort may make their effects felt for a full century or more.

The Antarctic region is valuable to the people of Earth because of its ecological role and because it is a mine of information useful to science. In addition, we have a responsibility to try to protect and to preserve, in as pristine a state as possible, one of the few areas on Earth not yet spoiled by man.

Opening up the continent for the pleasure of tourists will only make things worse. A continual infusion of sightseers will only make accidents like this oil spill more likely.

The Argentinian ship that went aground and spilled the diesel fuel was a supply ship for the Argentinian scientific station on the continent, but it also carried about a hundred tourists. Such ships should limit themselves to their essential supply functions and let the tourists find somewhere else to spend their leisure time.

Wetter and Warmer

It is never easy to proclaim flatly which of many possibilities might be considered "the most important scientific advance" of any period, particularly in the realm of actual discoveries, because the importance of some findings may not be immediately apparent. Back in 1944, a Canadian-American physician, Oswald Theodore Avery, discovered it was deoxyribonucleic acid (DNA), not protein, that carried genetic information. This was easily the most important discovery of the year and was well worth a Nobel Prize.

At the time, scientists weren't convinced of the importance of the discovery, and by the time it became clear that it had been a turning point in genetics and had led to a revolution in the field, it was too late to honor Avery appropriately. He was dead.

Therefore, as my candidate for the scientific event of recent times, let me choose something sociological, something that involves people rather than discoveries. The year 1988 was the time people became aware of something called the "greenhouse effect."

Adequate temperature records go back only to the 1850s, but 1987 was the warmest year for the Earth as a whole since then, and 1988 was warmer still.

Why is this? Because carbon dioxide in the atmosphere acts as a heat trap. Sunlight reaches Earth by day, passing through the atmosphere with little interference and warming Earth's surface as a result. At night, the Earth radiates that heat into space, such radiation taking the form of infrared waves. The chief components of Earth's atmosphere, oxygen and nitrogen, are as permeable to infrared as they are to ordinary light. Carbon dioxide, however, absorbs infrared and reradiates it in all directions. Some returns to the surface and keeps the Earth a little warmer than it would be if there were no carbon dioxide in the atmosphere.

This is a good thing. If there were no carbon dioxide in the atmosphere, the Earth would be in a perpetual ice age. Also, plants need carbon dioxide for photosynthesis. Without that gas in the atmosphere, no plants would grow and there would be no life on Earth, except, possibly, bacteria. If, however, there were more carbon dioxide in the atmosphere than there actually is, the Earth might become *too* warm.

The amount of carbon dioxide in the atmosphere before our present industrial age began was about 0.027 percent. This is very little, but it was enough to keep the plant world growing and to keep the Earth reasonably warm. Since that time, the amount of carbon dioxide in the atmosphere has been rising steadily. By 1958 carbon dioxide made up 0.030 percent of the atmosphere, and by 1988 it was nearly 0.035 percent—and rising. This increase doesn't seem like much, but it's enough to keep the Earth growing distinctly warmer.

The Earth's average temperature (evened out for changes of day and night, winter and summer) was about 58 degrees Fahrenheit (F) (14.5 C) back in 1880. It is about 59.5 F (15.4 C) now. That's a rise of 1.5 F (0.85 C) and that, too, doesn't seem like much. This rise, however, means a lot in terms of longer and warmer heat waves, longer and worse droughts—and, worse yet, rising sea levels.

Partly, the rising sea level arises from the fact that water expands as the temperature rises. There's a lot of water in the ocean, and even a 1.5 F rise expands it quite a bit. Since 1900, the sea level has risen by about six inches and it is still going up. Then, too, the warmer temperatures should cause the ice caps in Greenland and Antarctica to begin to melt.

If the ice caps melted completely (this would take a while, of course) the water formed would empty into the sea, and the sea level would rise by about two hundred feet. Places like the Netherlands, Bangladesh, Delaware, and Florida would be completely underwater.

There's even the possibility of a vicious cycle. As the water grows warmer, its ability to dissolve carbon dioxide grows smaller. That means that some of the carbon dioxide it now holds in solution will be released and find its way into the atmosphere, where it will act to warm the Earth still further.

This is not a discovery that was suddenly made in 1988. Scientists have thought about the greenhouse effect and been concerned about it for years. I myself wrote an article that was published in a magazine in August 1979 in which I said much of what I've said here. In other words, more than ten years ago, I was sounding the alarm, but, of course, no one listened.

Now, because of the heat and drought of 1988, the expression "greenhouse effect" has become familiar and people are listening. However, temperatures go up and down irregularly, and the next couple of years are apt to be a little cooler than 1988 even though the general trend is upward. If this happens, I imagine that people will forget again, until a year comes in the near future that is even worse than 1988.

But what can we do about the problem? To begin with, we must cut down on the burning of coal and oil. Such burning constantly pours carbon dioxide into the atmosphere (along with pollutants such as sulfur and nitrogen compounds that also act as heat traps and that are dangerous to the lungs as well).

Natural gas, when burned, produces less carbon dioxide than coal and oil do, and hydrogen, when used as a fuel, produces no carbon dioxide at all. Even nuclear energy is to be preferred to coal and oil, in my opinion. Nuclear energy has its dangers, but with a great deal of care it might be made safe and the radioactive wastes might be disposed of safely, too. There is, on the other hand, no way of making the burning of coal and oil safe.

The best source of energy, of course, would be solar energy. Light and heat from the Sun reach Earth anyway, and if we use this heat before it is absorbed by Earth's surface, this would in no way increase the heat content of Earth or change the atmosphere.

Working from the other end, how can we remove the carbon dioxide once it is in the atmosphere? The best way is to encourage the growth of forests. Trees absorb carbon dioxide and produce the all-essential gas oxygen more efficiently than any other form of vegetation.

And yet far from doing this, we are doing the opposite. The forests of the Earth are being chopped down at a huge rate, especially in tropical countries such as Brazil. This must be stopped. It is suicidal to maintain that what a country does within its own borders is its own business. The loss of the forests will increase the carbon dioxide supply and decrease the oxygen supply for all humanity. This is an example of the global nature of the dangers that oppress us and proves that no nation can be allowed to be a law unto itself. Solutions must be global, too.

Leap Second

The year 1987 was not a leap year. It consisted of exactly 365 days of 86,400 seconds each. Therefore, the total number of seconds in 1987 should have been 31,536,000. But it wasn't. The number of seconds was 31,536,001. The reason is that just before the year ended, an additional second (a *leap second*) was added by international agreement. Why? Because the Earth turns irregularly.

The irregular turning of the Earth is a comparatively recent discovery. The Earth is so massive that it takes huge forces to alter its rate of rotation; because its rotation apparently is so steady, it seems reasonable to predict that the Earth will make a complete turn relative to the Sun in 86,400 seconds, not a second longer, and not a second shorter.

The Earth's rotation, however, isn't perfect. There are shiftings of mass in the Earth. The liquid central core sloshes a bit as the Earth turns. There are earthquakes that redistribute masses here and there. During winter, masses of water are withdrawn from the ocean and deposited on land as snow, then in the spring the water is returned. There are water currents in the ocean and winds in the atmosphere. All this introduces wobbles in the Earth's turning so that it is a fraction of a second slow or a fraction of a second fast now and then.

This wouldn't be so bad because in the long run it would all even out, and even if you were off a fraction of a second now and then, it would never get worse than that. There is, however, one change that is cumulative. It involves the tides.

The Moon's gravitational effect is greater on the side of Earth that faces it than on the side turned away from it, because the side facing away is eight thousand miles farther from it. The result is a slight stretching of the Earth in the direction of the Moon. The stretching has a greater effect on the liquid ocean than on the solid land. There is a small bulge in the ocean on the side toward the Moon and another on the side away from the Moon. The Earth turns through those bulges so that the sea level creeps up the shore, then down the shore, twice a day. These are tides.

As the Earth turns, the two bulges of water scrape against the shallower sea bottoms (such as those of the Bering Sea and the Irish Sea) and also against the shores. This scraping produces friction, and, like all friction, it turns energy into heat. This means that the Earth is steadily losing rotational energy and its rate of rotation is slowing.

Even the tides can't affect rotation rate enough for us to notice. Astronomers notice it, though. If they study the positions of stars as reported by earlier astronomers, they notice that their positions shift steadily with time. When the Earth's rotation rate slows very slightly, the stars seem to lag behind and not reach zenith as quickly. The change is cumulative; after a couple of thousand years, the stars are way out of place even though the day hasn't lengthened noticeably. This is also true of eclipses. If we calculate where a solar eclipse should have been seen a few centuries ago, we find that the eclipse was observed—but miles away.

Studying past astronomical records just gives us a notion of how the rotation rate slows down on the average. How do we measure the rotation rate of Earth from day to day? For this we need a clock that moves more steadily than the Earth does. Such a clock was not developed until 1955.

By that time we had atomic clocks capable of counting the vibrations of atoms. You could count, say, 9,192,631,770 vibrations per second, and this would always stay the same; it would never be one vibration more or one less. If you then measure the length of each day (or the time between successive appearances of a particular star at zenith), you can tell that the day varies by a few vibrations from day to day, sometimes speeding up, sometimes slowing down, but in the long run, slowing down.

When the Earth's rotation has fallen behind by 0.9 seconds, a leap second is added and the Earth is in sync again. When this system was started in 1972, 10 seconds was added to bring the Earth back into sync. In the fifteen years since, 13 leap seconds have had to be added. They are added either at the end of June or at the end of December. This is something that is necessary not only for astronomers but for those who navigate ships, for those in charge of the world's radio and television communication, and so on.

Eventually, in the distant future we will have to add a leap second every single day, at which point we'll just have to agree that the day has become a second longer. We can then define the second as a very tiny bit longer than it is now so as to keep the same number of seconds per day.

A Map Too Good to Be True

In the early 1960s, something called the Vinland map was uncovered. It seemed to be a map of the North Atlantic as drawn from Scandinavian discoveries between A.D. 800 and

1100. Presumably, it dated from before the great age of exploration that began in 1400.

On the right-hand side of the map is the west European coast, including a quite recognizable Great Britain and Ireland, together with France, Spain, and, at the top, Scandinavia. In the mid-Atlantic to the west of France and Spain were a group of islands that presumably represent the Azores.

Then, in the North Atlantic, west of Scandinavia, are first Iceland then Greenland. Most interesting of all, to the west of Greenland, there is drawn a large island that must represent that portion of North America explored by the Vikings, that was known to them as "Vinland." Vinland shows two large inlets, the northern one ending in an inland sea and seeming to represent Hudson Bay, the southern seeming to represent the Gulf of St. Lawrence.

This map really does not affect our notions of the discovery of North America. The true discovery of the American continents was made more than 25,000 years ago, in the depths of the Ice Age, by Siberian hunters who followed the mammoth herds into what is now Alaska. Undoubtedly, we shall never learn the details of this momentous expansion of the human range.

Nor does it affect our notions concerning the importance of the work of Christopher Columbus. Columbus's sea voyage not only touched the American continents but led to their colonization by Europeans and to the entrance of those continents into the mainstream of human history. The earlier Viking discoveries, in comparison, led to nothing and were only a historical footnote. Columbus's was the effective discovery.

Nevertheless, the Vinland map gave us an excitingly better picture than anything that had existed earlier about the extent of Viking discoveries.

The question, though, was whether the map was authentic—or a forgery.

In 1974, chemists removed tiny particles of the ink from the map and subjected it to careful analysis. They detected the presence of titanium oxide. This is a perfectly normal component of inks, but only of modern inks. It was unknown and unused in late medieval and early modern times. On the basis of that finding, the Vinland map was declared a forgery and was retired from scholarly consideration.

In 1987, however, the map was subjected to analysis by still more modern methods involving bombardment with protons in a very narrow beam. The protons would be absorbed and scattered in different ways by different elements. Using this method, researchers did not find titanium, casting doubt on the earlier conclusion.

The absence of titanium does not by itself prove that the map is authentic, because it could have been forged with ink that did not contain titanium. Nor would matters be helped if the map were tested by such dating methods as carbon-14 analysis. That would give us only the age of the parchment. It might very well be that the parchment is ancient, and the drawing upon it of twentieth-century origin.

But why should scholars be so skeptical? If there is no definite proof that the ink is modern, then why not suppose that a Scandinavian geographer, at some time between 1100 and 1400, collected the reports of Viking sea captains and, from their descriptions, drew the map?

The trouble is that the drawing of Greenland, as pictured on the Vinland map, is too good. In 982, the Viking Eric the Red discovered Greenland, and Viking colonies were established on the southwestern coast. These survived, just barely, till nearly 1400, and then Greenland was forgotten. It was rediscovered by the English explorer Martin Frobisher in 1578. Again, it was just the southern tip that was known.

It was not until the late nineteenth century that the northern coasts of Greenland were explored. In 1892 the American explorer Robert E. Peary (who was later the first to reach the North Pole) explored the northernmost coasts of Greenland and established that it is an island. This was done only with the greatest difficulty.

And yet, on the Vinland map, presumably drawn at least five centuries before Peary, Greenland is shown as an island—and not only that, an island with roughly its correct shape. Even the Hayes Peninsula, in the far northeast (where the Thule air base now is located), is shown with reasonable accuracy.

Modern scholars are quite certain that the Viking seamen, skillful though they were, could not have circumnavigated Greenland, amid the polar ice and climatic rigors, with only the ships they had. Nor could they have determined polar latitudes if they had done so.

In other words, regardless of ink, parchment, or anything else, Greenland seems too well drawn for the Vinland map to be authentic.

The Mislaid Island

Sometimes very famous places are mislaid and people have to look for them carefully. Sometimes they find them, and sometimes they don't. There is an island, extremely important to American history, that has been mislaid and is still being sought.

It may seem impossible to mislay a place but it happens all the time. For instance, the Bible says that Noah's Ark finally made a landfall "upon the mountains of Ararat." Ararat is an ancient kingdom known to the Assyrians as Urartu, and we know where it was and where its mountains still are. What we don't know is to which *particular* mountain the Bible may have referred. There is a mountain we call Mount Ararat, but that's just a guess, even though people sometimes look for the Ark there.

Then there's the city of Troy, destroyed by the Greeks after a famous ten-year siege. It was somewhere in the northwestern tip of Asia Minor, but for many centuries people wondered exactly where it was, and even whether it existed at all. Finally, a German archaeologist, Heinrich Schliemann, thought he had found it, and it's generally accepted that he did, but we can't possibly be absolutely certain.

One of the most important battles fought in Roman history was the Battle of Zama, 202 B.C., in which the Roman Scipio beat the Carthaginian Hannibal, at last. It was the victorious end of a war that the Romans had nearly lost, so you'd think they would keep track of Zama and erect monuments there. They didn't, and to this day, although we know when the Battle of Zama was fought and what happened, we don't know exactly where Zama is.

But what about the island in American history? Well, on August 3, 1492, Christopher Columbus left Spain with three ships on the most famous voyage in history. He sailed westward for seven weeks and then on October 12, 1492, reached land somewhere among the Bahama Islands.

The island he reached was inhabited by people he called "Indians" (because he thought he had reached "the Indies"; that is, eastern Asia). The Indians called the island "Guanahani," or at least that's what the name sounded like to Spanish ears, but Columbus paid no attention to that. In those days, and for many years afterward, "natives" didn't matter, and what they called things didn't count. Columbus named the island "San Salvador" (meaning "Holy Savior"), took possession of it in the name of Spain, and then went on to discover other islands and make other voyages.

Columbus became a great American hero eventually, and we celebrate Columbus Day every October 12 (or the nearest Monday to it so as to make it a three-day weekend). On October 12, 1992, we will celebrate the five hundredth anniversary of his landing on Guanahani, and we should make a grand job of it, but the curious thing is that we don't know on exactly which island Columbus landed.

For a long time, in fact, there was no island in the Bahamas that was known as either Guanahani or San Salvador. There was, however, an island called Watling's Island, after an English pirate, John Watling. It is about sixty square miles in area (nearly three times as large as Manhattan). Because it lies well to the east of the island group in general, it seemed possible that Columbus had come upon it first. It was therefore renamed San Salvador, and it is now officially considered to be the island on which Columbus landed.

But is it really? Well, we might try to follow in Columbus's wake. He kept a very meticulous log of his voyage, giving winds, currents, distance covered, and so on. That log, unfortunately, is lost but part of a copy still exists.

Two oceanographers at Woods Hole in Massachusetts, Philip Richardson and Roger Goldsmith, have tried to reconstruct the voyage, using what remains of the log and also using the best knowledge we have of winds and currents. Knowing the speed of the ships and the direction from which they left the Canary Islands, a kind of dead reckoning could be calculated to estimate where the ship would have been on the very early morning of October 12.

There had been earlier attempts at this in which matters were adjusted to make the voyage end at San Salvador.

An attempt in 1986, without adjustments, ended three hundred miles too far west, because the estimates of speeds and currents and winds were wrong. One thing that even Richardson and Goldsmith weren't sure of was Columbus's compass. Columbus recorded its readings, but the direction in which the compass needle points, as measured at a specific spot on Earth's surface, varies from year to year, and we don't know exactly what the direction would have been in 1492 at the different places Columbus passed.

Even so, the dead reckoning ends at a point about fifteen miles south of San Salvador. This makes San Salvador look pretty good. However, a small island called Samana Cay exists about forty miles southeast of the calculated end point. It remains just possible that it may be the island Columbus reached. The chances are, though, that barring the invention of a time machine, we'll never know for certain.

When the Earth Was Too Hot and Too Cold

If you think that the climate here and there on Earth is not too good, there are times, every once in a while, when the climate is much worse, when parts of the Earth's land surface become absolutely uninhabitable.

The reason for this arises from the difference between land and water. Water has a higher heat capacity than land has. That is, a given amount of heat is absorbed by water with less of a temperature rise than is true for land subjected to the same amount of heat. Similarly, if it gets cold, water gives up heat, dropping to a low temperature, but land, giving up the same amount of heat, drops to a considerably lower temperature.

The result is that the ocean is cooler than the neighboring land in hot weather, and warmer than the neighboring land in cool weather. The ocean, when nearby, therefore exerts a moderating influence on the land's temperature so that the "oceanic climate" of coastal areas and of islands tends to be milder than it would otherwise be.

On the other hand, land that is far from the ocean does not have a chance to have its temperature moderated. It gets good and hot in the summer and good and cold in the winter. Such regions experience a *continental climate.*

Ordinarily, you would expect the North Pole and the South Pole to be the coldest regions on Earth by the time they haven't had any sunshine for six months. As far as the South Pole is concerned, that's almost correct because the South Pole is located on a continent. Still, the lowest temperature is not at the South Pole itself, but in that portion of Antarctica that is farthest from the ocean. The temperature there has been observed to be as low as −128.6 F (−54 C).

In the Arctic region, neither the North Pole nor any place near it holds the record for cold. The North Pole is in the center of the Arctic Ocean, where water moderates the temperature. The coldest region in the north is in central Siberia, far from the ocean and barely at the Arctic Circle.

The town of Verkhoyansk in Siberia has experienced a temperature as low as −94 F (−34.7 C) in the depth of winter. On the other hand, in the height of summer, that same town can experience temperatures of up to 98 F (37 C). That means there is a total range of up to 192 Fahrenheit degrees (89.6 C) of temperature because of the absence of any oceanic effect. (In the United States, it is places such as

North Dakota that get really cold in winter and really hot in summer.)

But the continents have not always been distributed as they are now. Very slowly, they are pushed here and there by the movement of the huge plates that make up the Earth's crust. Every once in a long while, they are pushed together to make up one huge supercontinent, called *Pangaea* (from Greek words meaning "all-Earth"). The last time this happened was about 255 million years ago, when the early reptiles (the ancestors of the dinosaurs and of us) were clumping the Earth.

Imagine Pangaea! It was three times as large as Asia and all in one piece. The central portions of Pangaea were perhaps two thousand miles farther from the ocean than any land surface can be today.

The central regions of Pangaea, if far enough north or south, would get colder than any place on Earth today in the deep winter and hotter than any place on Earth today in the height of summer.

Some scientists affiliated with the Applied Research Corporation and headed by Thomas Crowley have created a model of Pangaea's climate on a computer and reported on the results in the spring of 1989.

As might be expected, the climates in the interior would be ferocious. Summer temperatures may have routinely reached a high of 115 F (46 C) or even more, and the winters would have been way below zero. If one plots out the portions of Pangaea that would have had temperatures as bad as, or worse than, that of the north-central portions of Siberia and Canada, it turns out that these superuncomfortable regions of Pangaea encompassed at least eight times the areas of such regions on Earth today.

The places in Pangaea that would have been subjected to the highest temperatures were in what are now eastern Brazil and west-central Africa. The places where the greatest temperature range between summer and winter took place were in what is now southern Africa.

Fossil finds are few in those areas; the climate was probably so ferocious in central Pangaea that life simply could not stand it. This is especially likely because central Pangaea would have been so far from the ocean that the rains could rarely reach it no matter which wind was blowing, making it too hot and dry to sustain life.

It was fortunate, then, that Pangaea broke up (as it always had to, sooner or later). The fragments of Pangaea are all more moderate in climate than Pangaea itself was, and the rims of the oceans near the land are always richest in life, too, and with many continents there is more ocean rim than there was with just one supercontinent. So life today is better off—if only we don't do anything to ruin things.

Ice Ages and the Plateau Effect

One of the perennial mysteries of Earth's history is the reason for the ice ages, for the coming and going of the vast glaciers. In spring 1989, a possible solution to the mystery was advanced by William P. Ruddiman of Columbia University and John E. Kutzbach of the University of Wisconsin.

As long ago as 1920, a Yugoslav physicist, Milutin Milankovich, pointed out the astronomical facts of the case.

Earth's orbit changes slightly in cycles. There are slow increases and decreases in the tipping of the axis, in the degree to which Earth's orbit is not quite circular, in the position of Earth's perihelion (its closest approach to the Sun), and so on.

The net result of all these variations is a slow and slight increase and decrease of the amount of radiation we get from the Sun in a cycle that lasts about 40,000 years. Every 40,000 years, in other words, the Earth goes through a 10,000-year-long shortfall of radiation. Its average temperature drops a bit and there is a "Great Winter."

During this Great Winter, the abnormally cool summers don't suffice to melt all the snows of winter. From year to year the snow accumulates and the glaciers advance. With the passing of the Great Winter, the glaciers retreat again.

This makes sense, and the careful study of past fossils indicates that such a temperature cycle may indeed have existed. If it did, however, it must have existed for billions of years, but the ice ages have only been coming and going in the last million years or so. Before that, there was a period at least 250 million years long in which there were no ice ages.

It would seem that the Great Winter is ordinarily not cool enough to start an ice age. In the last couple of million years, something about the Earth must have changed to make the cool periods more effective. The changes can't be astronomical but must involve the Earth itself. Suspicion centers on the slow shifting of the tectonic plates that make up Earth's crust and the consequent shifting of the continents.

In 1953, two Columbia University geologists, Maurice Ewing and William L. Donn, pointed out that the shifting continents must have come to encircle the North Pole only a couple of million years ago, leaving the Arctic Ocean at the center. The Arctic Ocean was a source of moisture and the snow fell on the vast stretches of Canada and Siberia. Snow doesn't melt as rapidly from a land surface as from a water surface. The snow therefore accumulates more readily if land surfaces, instead of open sea, surround the North Pole. Therefore, it is only since the continents drifted into their present arrangement that the lower temperatures of the Great Winter could bring about ice ages in the Northern Hemisphere.

Now Ruddiman and Kutzbach have an alternate suggestion that seems particularly attractive to geologists. They point out that, as a result of the shifting plates, the landmass we call India, originally a large island, plowed slowly into the southern rim of the continent of Asia. The land at the point of contact slowly crumpled and lifted upward, forming the high Himalayan Mountain range and the vast Tibetan plateau.

Similarly, the continent of North America has been moving westward into the Pacific, and the frictional forces have crumpled its western regions into the Rocky Mountain chain.

Over the past 20 million years, portions of western North America have been raised a mile into the air. The Himalayan regions have been raised three miles into the air.

Before these particular changes, the landmasses of the Northern Hemisphere were relatively flat, and the winds could move more or less unobstructed from west to east around the world.

Ruddiman and Kutzbach worked out computer simulations of the wind pattern that would result as regions in central Asia and North America lifted into plateaus and mountain chains. They found that, because of the high regions, the winds tended to be deflected and to move farther north than they had moved before. The air masses, as a result of these northern deflections, cooled down and brought lower temperatures into the regions northeast of the Rockies and north of the Himalayas.

With lower temperatures, the amount of snow melting in the summer decreased and the onslaught of the Great Winter was more vigorous. Finally, in the last couple of million years, the plateaus and mountains had risen high enough, and the wind deflection had become pronounced enough, to bring about a cooling effect sufficient to produce a Northern Hemisphere ice age during the Great Winter.

If this is so, we can expect periodic future ice ages until the Rockies and Himalayas have been worn down sufficiently to become less effective as wind deflectors. Unless, of course, the human-made warming trend of the "greenhouse effect" puts an end to ice ages altogether, and brings on its own catastrophes of a different kind.

Moon Mysteries, Earth's History

Twenty years ago, on July 20, 1969, human beings set foot, for the first time, on a world other than Earth itself. Neil Armstrong stepped down onto the Moon's surface and said, "This is one small step for a man, one giant leap for mankind."

There were five more visits to the Moon in the next few years, and then they stopped. No one has visited the Moon now in seventeen years. So as we celebrate the twentieth anniversary of that first Moon landing, we might ask: What was it all for? Did we get anything out of it? Was there really a giant leap?

Well, yes; the Moon landings gave us a valuable chance to learn about its—and our—far past.

The Earth and the Moon, and the whole solar system, in fact, came into being about 4.6 billion years ago. We can learn about the Earth's distant past by studying its rock formations. Naturally, the older the rock and the longer it has existed in Earth's crust unchanged, the farther back our knowledge can be pushed.

However, the key word there is *unchanged.* Rocks don't remain unchanged forever. The Earth's crust shifts and rocks are crushed and melted and re-form. The force of moving air and water introduces changes even when the rocks aren't actually melted. And life itself changes the landscape enormously.

The result is that the oldest rocks we can find are a little more than 3 billion years old, and there aren't many of those. We have difficulty extending our knowledge of Earth's history back that far, and as for the first 1.5 billion years of Earth's existence, forget it. It remains a complete blank, and as long as we are imprisoned on Earth itself, it will continue a complete blank.

The Moon, however, is a smaller body. Its gravitational pull is insufficient to hold an atmosphere or any liquid that evaporates easily. This means that there is no air on the Moon and no water and no life. Not only is there none now, but there never was any. This, in turn, means that the surface has not been interfered with by the action of life or by wind or by waves. What's more, being a smaller body, the Moon has developed less internal heat, and it is internal heat that keeps the crust moving and changing even if nothing else does. In other words, whereas the Earth is geologically "alive," the Moon is virtually geologically "dead."

This means the surface of the Moon can continue to

exist unchanged much longer than the surface of the Earth can, and the Moon rocks that the astronauts brought back are a billion years older than the oldest rocks we can find on Earth's surface. We can fill in a billion years of early history that the Earth is silent about.

The Moon was created (it is currently thought) when, very early in Earth's history, the planet was clobbered by an object about the size of Mars. This knocked a huge quantity of Earth's surface layers into nearby space while the striking body fused with Earth.

The material that was knocked into space was heated to a vapor, but it cooled into a mass of innumerable particles of various sizes and these gradually coalesced and formed the Moon. Because the Moon was formed from the outer layers of Earth, it is almost entirely rocky and includes very little iron like that found at the Earth's core. This is why the Moon is less dense than Earth.

The Moon took a few hundred million years to cool down sufficiently to have a solid crust, but about 4 billion years ago this solid crust was there, and the oldest rocks that have been brought back date from that period.

In the last 4 billion years, the only significant changes undergone by the Moon happened when it gathered in the remaining objects that still existed in its vicinity. These formed the numerous craters and the vast "seas" that now cover its surface. From the rocks we have brought back, we can study the different stages of this bombardment. The early history was the most active, of course, because there were still many objects with which to collide.

As time went on, space was cleared of most of the objects and the Moon settled down and underwent fewer and fewer other changes. From about 3.2 billion years ago onward, things were relatively quiet—for Earth as well as the Moon, for if the Moon was bombarded, so was the Earth. It's just that on Earth the craters produced by that bombardment have worn away, thanks to winds, waves, and life, whereas on the Moon they remain.

Yet there have been changes on the Moon comparatively recently, too. The crater of Copernicus was formed 810 million years ago, and the spectacular crater of Tycho was formed only 109 million years ago. Some small craters were formed as recently as 2 million years ago.

If we were to go back to the Moon, then, quite apart from its uses as an observatory, a mining station, and a new home for human beings, a careful and painstaking study of its surface could fill in all the details of its history, and from that we could deduce what took place on the early Earth, too. The clues we find there might help us figure out how life began on this planet and how we all came to be.

IV

FRONTIERS OF SPACE

The Cracked Crust

The Earth is unique among the bodies of the solar system.

Suppose we omit the Sun and the four giant planets, Jupiter, Saturn, Uranus, and Neptune, all of which are made up largely of the gases hydrogen and helium. Aside from these gaseous worlds, all the bodies of the solar system—planets, satellites, comets, asteroids, meteoroids—are made up of icy, rocky, or metallic substances or some combination of these.

Of all these bodies, the Earth is the largest. It is the only body that is warm enough to have an open ocean of liquid water, but not so warm that the ocean boils away. It is the only body that has an atmosphere that contains free oxygen.

Of course, we've known about Earth's size, ocean, and air for a long time. There is, however, another point about Earth that may be unique, which we've only known about for a quarter of a century.

Because the Earth is the largest of the nongaseous worlds, it has a higher internal temperature than any other and therefore has the thinnest crust. What's more, because of the Earth's high internal temperature, there is a great deal of energy in there, so that the Earth can behave as a powerful heat engine, more than the other smaller, internally cooler, bodies of the solar system.

The result is that the Earth's thin crust is cracked into a half-dozen large pieces (and several smaller pieces) called *plates*. These plates fit tightly together as if they had been fitted by a clever carpenter. They are therefore called *tectonic plates* (*tectonic* is from a Greek word meaning "carpenter").

The rocky matter below the crust is hot enough to be able to move in slow swirls, and this movement pushes the plates this way and that. Some adjoining plates are pulled slowly apart, leaving a basin that fills with water so that an ocean is very slowly formed. The Atlantic Ocean was formed in this way over the past 200 million years.

Two plates may be pushed together and crumpled. In this way highlands and mountain ranges are formed. The Himalayan Mountains and the Tibetan plateau were formed when two colliding plates pushed India into Asia. Or else one plate may dive under another, forming ocean deeps. At the boundaries where the plates meet are lines of weakness, where volcanoes and earthquakes make themselves evident. The San Andreas fault may be the best known of these boundaries. Almost everything in the Earth sciences can be explained by these plates, but their existence wasn't discovered until the early 1960s.

Smaller bodies than the Earth have less heat inside and, therefore, thicker crusts. The crusts don't crack, so that they consist, so to speak, of one world-girdling plate. The Moon, Mercury, and Mars are all one-plate worlds and, therefore, "geologically dead," at least compared to the Earth. Mars, however, does have volcanoes. Though they seem to be extinct now, they must once have been alive.

On Earth we have lines of volcanoes. As the plates move, new volcanoes appear in new places, and we end up with something like the chain of volcanic islands that make up Hawaii. On Mars, however, without plate movements, the volcanoes build up in one place so they are monsters far larger than any we have on Earth. (The satellite Io has active volcanoes, but the energy is supplied by the tidal effect of giant nearby Jupiter.)

But what about Venus? Venus is smaller than the Earth but not by much. It is about 81.5 percent as massive as Earth is. Until about ten years ago, Venus's surface was hidden from us by an eternal layer of clouds. However, we are now capable of studying the surface by radar, which can penetrate the clouds.

Radar waves are much longer than light waves, so they don't show detail as sharply, but even the first radar studies in 1978 showed that Venus had large highland areas that resembled Earth continents and even larger lowland areas that looked as if they might once have contained oceans. Of course, since Venus's surface temperature is now far above the boiling point of water, any ocean it may have had must have boiled away billions of years ago.

In recent years, the Soviets (who have made Venus their specialty) have taken much better radar photographs. These show craters on Venus, which, from their appearance, seem to be anywhere from half a billion to a billion years old. This speaks against plate tectonics, for on Earth the plate movements keep renewing the surface. Sixty percent of Earth's surface is less than 200 million years old.

On the other hand, there are plenty of indications that there may be plate movements on Venus. There are mountains where plates may have come together and rifts where they may have pulled apart. What we need are still better images and closer studies, but it seems that Venus is an intermediate case. It may have tectonic plates, but they are probably nowhere near as active as ours.

So, all in all, Earth may still be viewed as a unique world.

Blast over Siberia

Scientists are still wondering about an incident that took place in central Siberia more than eighty years ago. They are still poking around in the area and, even today, are uncovering new evidence concerning it.

On June 30, 1908, the skies over central Siberia, near the Tunguska River, lit up and there was a terrific blast. Hundreds of square miles of forest were leveled—every tree down. A herd of reindeer was wiped out.

Fortunately, there were no people within many miles of the explosion in that desolate area so that no human being was killed. However, a person at a trading post fifty miles away was knocked out of his chair by the blast's force, and other distant observers saw, heard, and felt the effects.

It took a long time for scientists to make their way out to that all-but-inaccessible place, and things weren't helped by the fact that World War I came along, followed by years of revolution and civil war in Russia. It wasn't till well into the 1920s that Soviet investigators finally reached the scene.

That was when the real mystery started. Everyone assumed that a large meteorite had struck Siberia, one that weighed anywhere from 100,000 to several million tons. It might have been a lump of rock about 250 feet across or a lump of iron about 80 feet across. In either case, as it came speeding into the Earth at twenty miles a second or so, it would have done the damage of a large-size hydrogen bomb (minus the radioactive fallout, of course).

Such an impact would have gouged out a large crater and would have perhaps left the meteor buried in the Earth or strewed the landscape with pieces of iron or meteoric rock. Investigators found the exact point of impact, since all the fallen trees pointed away from it, but at that point there was neither a crater nor meteoric pieces.

The only reasonable conclusion was that the explosion had taken place not on impact with the ground but in the air, perhaps five miles above the ground. The incoming object never reached the ground, in other words, and simply distributed itself through the atmosphere. Indeed, the blast had produced waves in the atmosphere that had been detected everywhere in the world.

This would be a strange way for a meteor to react, however. Stone or metal would not explode in midair like that. But what if it were not stone or metal, and not an ordinary meteorite? It might have been a small comet, about three hundred feet across, or a fragment of a larger one.

A comet is made up largely of icy materials, frozen water chiefly. As it raced through the atmosphere, air resistance would raise its temperature. Rock or metal would begin to glow and we would see a "shooting star." Ice, however, would vaporize. If the comet grew hot enough quickly enough, the sudden vaporization could produce a huge explosion, shattering whatever portion of the comet had not yet had a chance to vaporize. The resulting gases (chiefly water vapor) would spread through the atmosphere. Nothing would reach the ground, except the blast, and there would be no crater and no meteoric fragments.

This seemed a completely satisfactory explanation. Of course, other explanations of the Tunguska incident were advanced. It might have been a tiny chunk of antimatter exploding when it reached the ordinary matter of Earth's crust, leaving nothing behind. Or it might have been a nuclear spaceship from some other world that exploded.

Such alternate explanations, however, were never taken seriously.

But in early 1987, a group of Soviet investigators reported that they had detected abnormally high traces of the metal iridium in soil obtained from the impact point. Iridium is rare in the Earth's crust, for most of it has settled down in the Earth's central core. It exists in considerably higher concentrations in meteorites, so the presence of higher than usual traces in the crust is considered a relic of a meteoric impact. Comets, however, do not contain any iridium to speak of. The new Soviet finding seems to indicate the Tunguska event was caused by a meteor, after all, and not a comet.

Then where's the crater? The Soviet investigators suggest that the invading object was a comet that was surrounded by dusty material rich in iridium. This would account for the combination of the presence of iridium and the absence of a crater. Not everyone is going to accept this, though. The event remains a puzzle that inspires more frightening conjecture.

Two points can be made. Central Siberia is about the only place such an event could have happened without human casualties. If it had happened at sea, it would have set off tidal waves. Almost anywhere else on land, people would have died, perhaps in vast numbers.

Second, suppose such an event happened now in the Soviet Union or in the United States. Such a blast might be taken to be an enemy nuclear bomb, and a vengeful counterstrike might be ordered at once. The horrors that would follow would be indescribable.

Halley's Comet

For astronomers, 1986 was the year of Halley's comet. The probes that were sent out in its direction managed to photograph a comet and study it at close quarters for the first time in history.

Why did we take the trouble, and what did we find out?

Astronomers are interested in the details of how the solar system came into being. This would help us understand how the Earth formed and how life came into existence. The trouble is that the only thing we have to work with are what we can find out about the Sun and the planets *now.*

All these bodies are 4.6 billion years old and they have undergone enormous changes. For instance, the bodies near to the Sun (such as the Earth itself) have been heated for billions of years and have lost the easily vaporized material that might have made up most of their original structure. If we studied only Earth, we could only guess at what it was like when it was first formed.

Bodies that are farther away from the Sun are likely to be less changed, but because of their distance, they are not so easily studied.

The farthest objects of all are the comets. A hundred billion or more of them circle the Sun slowly at distances of 1 or 2 light-years, thousands of times as far away from the Sun as the most distant ordinary planet. At such distances, we can't study them; we can't even see them; we can only suspect their existence from indirect evidence. However, every once in a while, the gravitational drag from nearer stars sends some of those comets into the inner solar system and the neighborhood of the Sun. They can then be studied.

Astronomers have done their best to work out the chemical composition of the original cloud of dust and gas from which the planets were formed. A comet is made up of that original dust and the cases that have frozen about those dust particles to form ice. When Halley's comet approaches the Sun, that ice vaporizes, and if the gases produced are analyzed, we will have samples of the original material out of which the solar system was formed.

These gases were indeed analyzed by the probes, and they turned out to have identities and to be present in proportions very close to what astronomers have suspected. This is great news. It is useful to deduce a logical possibility from indirect evidence, but it is much more useful to have the possibility confirmed by actual direct measurements. Astronomers can now work out the details of where we began with much greater confidence and can move forward boldly.

Halley's comet did not, however, only give us confirmation of matters that astronomers had suspected. It gave us at least one big surprise. It turned out to be dark black in color.

Back in 1951, Fred Whipple, the greatest comet astronomer alive, had presented his reasons for believing comets to be "dirty snowballs." That is, they were formed of icy materials—mostly ordinary water ice—with an intermixture of rocky dust and grit that made up the "dirtiness." The closeup studies of Halley's comet confirmed this. It is about five-sixths water ice. Other comets are, presumably, the same.

Naturally, when a comet approaches the Sun, some of this water ice (together with other icy materials) is vaporized and is gone, but much of the rocky dust and grit remains behind. The surface of the comet tends to be coated with a thicker and thicker layer of dust that darkens the comet.

Astronomers thought, therefore, that comets might be grayish, but they were not prepared for outright black.

Assuming a light-colored comet that reflected most of the light that fell on it, astronomers estimated that Halley's comet might be four miles across. However, Halley's comet is black and reflects very little light. To be as bright as it is, it must be much larger, and, in fact, closer measurements show that it is more like ten miles across. It contains twelve times as much material as astronomers had thought.

Presumably, then, comets generally are considerably larger than had been thought. The belt of comets that exists about 1 or 2 light-years from the Sun had been thought to have a total mass twice that of the Earth. Instead, it may be that the total mass of comet that exists far out from the Sun may be no less than twenty-five times the mass of the Earth.

Many astronomers believe that in the early times of the solar system, there were a large number of collisions between comets and the planetary bodies. With the new information we have at hand, it might seem that such collisions may have been if not more numerous then more massive and more individually devastating than had been thought.

The Earth was probably hot and dry to begin with, and cometary collisions may have supplied us with much of our ocean and atmosphere. That seems more likely now. It is also more likely that cometary collisions can produce periodic waves of extinction of life and that they killed off the dinosaurs. All this we can now reason out as a result of the close study of Halley's comet in 1986.

More About Halley's Comet

It turns out that the human species has been in this neighborhood of the galaxy considerably longer than Halley's comet has, according to three Canadian astronomers headed by J. Jones. They reached that conclusion early in 1989.

Of course, Halley's comet, like all comets, is as old as everything else in the solar system: 4.6 billion years old. Astronomers think that comets, by the hundreds of billions, exist in a belt that orbits the Sun far, far beyond Pluto. The comets are icy, made up largely of frozen water and gritty dust. Out beyond Pluto, however, the temperature is only a few degrees above absolute zero. There the comets last, unchanged, for many billions of years.

Every once in a while, though, something happens to send an occasional comet down into the inner solar system. Perhaps two of them collide and one drops sunward; or perhaps the gravitational pull of a passing star does the trick.

A comet that falls out of the cloud and into the inner solar system spends a portion of its orbit rather near the Sun. Large objects like Earth and Venus aren't bothered by being so close to the Sun. The Earth is composed mostly of metal and rock and isn't affected by the solar heat. Comets, though, are small and icy. The ice they're made of vaporizes. The dust content is liberated, forms a haze about the comet, and is swept back by the *solar wind* (streams of speeding charged particles emerging from the Sun) into a long tail.

It's a spectacular sight, when it is close enough to Earth

and high enough in the sky, but, of course, every bit of that vapor and dust has left the comet never to return. The next time that the comet, in its long orbit, returns to the neighborhood of the Sun, it is smaller than it was before—and loses more of its substance.

Astronomers have watched some small comets come to an end. Some have broken into two or more pieces that finally disappeared. Some have actually plunged into the Sun.

However, although the rocky grit never returns to the comet, it doesn't actually disappear into nothingness. It continues circling the Sun in a long cometary orbit. These are *meteor swarms*. Every once in a while the Earth passes through the orbit of such a swarm and numerous "shooting stars" flash across the sky. Once, in November 1833, they were so numerous over the New England sky, it appeared to be snowing. These bits of meteoric dust do no harm, however, and may indeed do good, for they serve as nuclei for raindrops and therefore encourage rainfall.

Earth's passage through the meteor swarms tells us where they're located and what their orbits must be like, and we can identify some of them with the comets from which they originated. There is a meteor swarm that originated in Halley's comet and that has spread out through portions of its orbit. Earth passes through it twice each year, once on one side of Earth's orbit and once on the other side.

The three Canadian astronomers have studied this meteor swarm by using a computer to create a model of what happens to crowds of simulated particles. From this, they have been able to deduce that the main body of the meteor swarm is 40 million miles long and 4 million miles wide.

From this, and from the number of particles in each little bit of the swarm, they reached the conclusion that the total material in the swarm is about 1.3 million tons. And every bit of that tonnage came originally from Halley's comet.

We don't know how much mass Halley's comet has, not for sure, but from the observations made by a probe on the comet's last trip through the inner solar system in 1986, it might seem that the dust in the meteor swarms represents one-tenth the total mass of the comet. What's more, the comet is mostly ice, and the water vapor doesn't show up in the meteor swarms. If we consider the loss of vapor as well as grit, then the mass of the meteor swarm makes it seem that about one-quarter to one-third of the original Halley's comet has by now disappeared.

Once this is decided (and the figures are, of course, only very approximate), it can be calculated how much mass Halley's comet must lose every pass it makes near the Sun and, therefore, how many passes it has made since it was trapped, somehow, in its present orbit.

The Canadian astronomers have calculated that Halley's comet dropped out of its distant cloud about 23,000 years ago, and in that time has completed some three hundred turns about its 76-year orbit. Three hundred times, human beings might possibly have gazed up at that same comet in the sky (on some occasions more spectacular-looking than on others, of course). *Homo sapiens*, however, has probably existed on Earth for some 50,000 years, and this means that for more than half our life span on Earth so far, no one ever saw Halley's comet (though, of course, other comets undoubtedly lit up the sky).

What's more, Halley's comet continues to lose mass and very likely loses a larger percentage of itself with each pass as it gets smaller. Halley's comet may, therefore, not last another three hundred passes, and, naturally, it will grow less and less spectacular. Other comets will come our way, but humanity will lose its young neighbor.

The Largest Molecule

When scientists had the once-in-seventy-five-years chance to observe Halley's comet up close via space probes in 1986, a great deal was added to our knowledge of what the comet is made of and what it actually looks like. One of the things the probes discovered may actually help solve one of the riddles of how life on Earth developed.

The carbon atom is central to life as we know it. Living tissue is made up of large, complex carbon-containing molecules (that is, atom combinations) such as those of proteins and nucleic acids. It has always been assumed that Earth in its earliest stages contained only very simple carbon-containing molecules, such as methane (one carbon atom and four hydrogen atoms) and carbon dioxide (one carbon atom and two oxygen atoms). The problem has been to work out how the large complex molecules that now exist formed from the simple ones that originally existed. No completely satisfactory solution has yet been found.

But how sure are we of the starting materials? The Earth (along with the Sun and all the other planets) formed, about 4.6 billion years ago, from a vast dust cloud, and we're not quite certain of what that dust cloud consisted. Mostly, scientists are sure, it was made up of hydrogen and helium, for this is what the Sun and the giant planets are composed of. Still, it must also have had small amounts of carbon atoms, or Earth would have no carbon and, therefore, no life. In what form did those carbon atoms exist?

There are many dust clouds in space, some of them in

the process of forming stars. Of what materials are those dust clouds made?

There was no hope of getting an answer to such a question until about a quarter of a century ago, when radio telescopes had become sufficiently sophisticated. Each type of molecule radiates radio waves of certain lengths. These act as a kind of fingerprint that can be detected by radio telescopes.

Atoms are so thinly spread out in the space between stars, even in the gas clouds, that astronomers thought collisions would be rare. Therefore, they thought that any molecules that existed there would consist of, at most, two atoms. In 1968, however, much to their astonishment, astronomers discovered the telltale radio-wave indications of the presence of molecules of water (made up of three atoms) and ammonia (made up of four atoms).

In fact, by now, they have discovered dozens of different molecules in gas clouds, some of them too unstable to exist on Earth. Some of these molecules consist of as many as thirteen atoms! How all these atoms could get together when they are so thinly spread out is still a matter of considerable argument.

One important point, however, is that every molecule of more than four atoms contains one or more carbon atoms. In space, as in our bodies, complicated molecules are composed of carbon atoms.

But might not there be still more complicated carbon-containing molecules in dust clouds than any we've so far located? This seems very likely. The more complicated a molecule, the rarer it is and the more difficult it would be to detect. The really complicated ones would simply not have been detected yet. In fact, they may never be detected, since the gas clouds are so far away. But something closer might provide a clue.

When our own gas cloud formed our solar system, all

the resultant objects contained any complex molecules that may have been present in the cloud. However, in the course of the development of large bodies, such molecules would be broken up by heat and other factors.

On the very outskirts of the cloud, however, most of the matter must have clumped together into billions of small bits of icy material in bodies of only a few miles across. In such small objects, so many billions of miles from the developing Sun, the complex molecules might be preserved indefinitely. Occasionally one of these distant bits of matter wanders into the inner solar system and the heat of the Sun vaporizes parts of it. It then becomes visible as a comet.

The dust and gases that surround a comet as it passes through our own neighborhood may contain interesting molecules for this reason. The rocket probes, especially the European probe named Giotto that passed the closest to Halley's comet, tested for this.

Now, it has been reported by Walter E. Huebner of Los Alamos National Laboratory in New Mexico that Giotto detected in Halley's comet a *polymer,* a combination of formaldehyde molecules (long known to exist in space) in an indefinitely long chain. Such chains may help account for the surprising darkness of the comet surface, and they may have existed in the dust cloud out of which the comet (and Earth) originally formed.

May it not be, then, that as the Earth formed, some complex molecules escaped destruction and persisted in isolated spots? If this were so, then life need not necessarily have begun to form from simple carbon compounds, all the way from the beginning, but may have had a boost. Some of the complex molecules necessary for life may already have existed in the dust cloud at the time of the Earth's formation. If so, this could make the process of the origins of life more understandable.

Our Identical Twin

In January 1989, some new deductions were made about the nature of volcanoes on Venus. These findings may eventually help shed some light on our own planet Earth.

Until about thirty-five years ago, nothing was known—nothing!—about the details of the structure of other worlds, not even of our own Moon. Since then, thanks to radar astronomy and rockets, we have learned a good deal, but the planet we are still most curious about is Venus.

The reason for this is that Venus is, in some ways, more like Earth than any other planet is. Whereas Earth has a diameter of 7,926 miles, that of Venus is 7,550 miles. Venus has about 81 percent of the mass of the Earth and 94 percent of its density. Venus has an Earth-like structure, with a rocky outer layer and a liquid metallic core. It is almost an identical twin, except—

Earth rotates on its axis from west to east in twenty-four hours; Venus rotates from east to west in 244 days. Venus has an atmosphere that is about ninety times as dense as Earth's and one that is about 95 percent carbon dioxide and no oxygen, whereas ours is only 0.03 percent carbon dioxide and 21 percent oxygen. Earth's surface temperature is about 300 degrees above absolute zero, and Venus's is more than 700 degrees above absolute zero, hot enough to melt lead. Venus has no water on its surface; Earth has oceans of it.

And, of course, Earth is rich in life and Venus has none at all. If we could figure out why two planets should seem identical twins in some ways and yet be so unidentical in others, we might understand a great deal more about both Venus and Earth.

One of the things we know about Venus is the nature of its rocky surface. Soviet rockets have landed on its inhospitable surface a number of times and found that one of the main chemicals composing the surface is calcium carbonate. This is not too surprising. When hot, calcium carbonate tends to break up and liberate carbon dioxide, so a hot planet with lots of calcium carbonate in its surface rock is sure to have lots of carbon dioxide in its atmosphere, as Venus does.

Another thing we know about Venus is that its clouds do not consist of droplets of pure water as on Earth. They consist of droplets of corrosive sulfuric acid.

However, Ronald Prinn of Massachusetts Institute of Technology pointed out that limestone will combine with sulfuric acid to form calcium sulfate and carbon monoxide. He and a colleague, Bruce Fegley, ran experiments in early 1989 to find out how rapidly calcium carbonate and sulfuric acid would react at the high surface temperature of Venus and how long it would take such a reaction to clear the sulfuric acid out of Venus's atmosphere.

Their results and calculations led them to believe that Venus's atmosphere ought to lose all its sulfuric acid in 2 million years. This is a long time in comparison with human lifetimes, but it is almost nothing to a planetary lifetime, because Venus (like Earth) is about 4.6 billion years old. Venus's sulfuric acid should have been gone long, long ago.

But the sulfuric acid has *not* been eliminated. It is still there. This means that new sulfuric acid must be forming as fast as old sulfuric acid is removed. The most likely way in which new sulfuric acid is formed is by volcanic action.

Prinn and Fegley then calculated how much volcanic action was needed on Venus to keep the sulfuric acid of the atmosphere constant. It turned out that an amount of volcanic action on Venus just 5 percent of the rate of volcanic action on Earth would do the job.

This is supported by another point. If Venus were as volcanic as Earth is, most of the craters on its surface would be filled in and covered up by lava flows. The craters, however, are *not* filled in, and again the calculation would make Venus only about 5 percent as volcanic as Earth is.

But this presents a new puzzle. Earth has a crust that is rather thin and is split into plates that slowly move about. The internal heat of the Earth can leak out at the joints the plates make with each other and also by way of volcanoes.

Venus has a thicker crust not broken into plates. Its internal heat can escape only by way of volcanic action. If we suppose that Earth and Venus, being of nearly equal size, should have the same amount of internal heat and should lose it at the same rate, then Venus ought to be much more volcanic than Earth, perhaps a hundred times as volcanic because there are no plate joints from which it can lose heat.

But this is impossible on the basis of the new studies, so planetary scientists have to conclude that Venus has a much lower internal heat than Earth has, or that it has some way of losing heat that doesn't involve either plate joints or volcanoes. Neither possibility seems to be at all likely, but it's such puzzles that make science exciting and that hold the promise of a brand-new outlook on matters—including, possibly, Earth's own structure—once they are solved.

Microwaves as Cloud-Zappers

We are accustomed to supposing that if we want to see a distant astronomical object in detail, we must send out a probe. But that's not necessarily so. We can see much detail from right here on Earth, and sometimes find out in this way more than a probe can tell us. We can reach out from Earth and touch another world, as we have now done in the case of Saturn's large satellite, Titan.

We did get a close look at Titan when Voyager 2 passed by Saturn a few years ago. It's a large satellite, about 3,250 miles in diameter, which makes it considerably larger than our Moon. It has an atmosphere, too—which our Moon does not.

Titan's atmosphere is thick and is made up of nitrogen and methane for the most part. Nitrogen is unaffected by sunlight. Methane, however, has a small molecule made up of one carbon atom and four hydrogen atoms, and the energy of sunlight knits these molecules together into larger units. These larger units are hydrocarbon molecules, like those in gasoline.

Even though Titan is nearly ten times as far from the Sun as Earth is, there is enough sunlight reaching the satellite to form these larger molecules. The result is that Titan's atmosphere seems to be full of gasoline vapors that form a hazy smog. The instruments on Voyager 2 could not penetrate that smog; all they could send back were photographs of a hazy circle of dim light.

Naturally, scientists are curious about what the solid

surface of Titan might be like. Is it solid? Is it covered with an ocean of liquid hydrocarbons, or liquid nitrogen, or a mixture of both? Or what?

Finding the answer may seem hopeless until we can return to Titan and actually send a probe into its atmosphere and down to its surface.

However, scientists had a similar problem with an object much closer to us than Titan is: the planet, Venus. Venus is always covered with a thick layer of clouds that our telescopes cannot penetrate. There seemed to be no way of telling what the solid surface was like. We couldn't even tell if Venus was rotating, or, if it was, in what direction, and how fast.

However, a beam of microwaves, such as those produced by radar instruments, can penetrate the clouds. The beam can reach the surface of Venus, bounce back, penetrate the clouds again, and come back to us. Scientists can then detect this "microwave echo."

If Venus's surface were smooth and motionless, the microwave echo would be just the same as the beam that was sent out in the first place. If the surface were turning, however, the echo would come back with a different wavelength. From the change in wavelength, scientists would be able to tell how fast Venus was turning and in which direction.

If Venus's surface is uneven, the microwave echo is scattered a bit, and from the scattering you can tell just what the unevenness is like. In fact, we have now produced maps of Venus's surface as revealed by microwaves.

The question is, can we do the same for other worlds? Clearly we can, for we have sent out microwave beams that have bounced off Mars, Jupiter, Mercury, even the Sun, and have detected the echoes.

Can we bounce microwaves off Titan? Yes, but even at its closest, Titan is about thirty-five times as far away as

Venus is from Earth. This means that a microwave beam will produce an echo from Titan that is only about one-twelve-hundredth as strong as an echo from Venus. We would need a stronger beam to begin with, and instruments that can detect weaker echoes.

Such instruments have been developed. The new signal-detecting instruments have been used for Titan. In June 1989, microwave beams were sent to Titan and echoes were detected—the weakest echoes scientists had ever worked with.

The beam was sent out on three different days, June 3, 4, and 5. Since Titan is turning, with each turn taking sixteen days, the microwave beams hit a different part of the surface each day. (It would be as if we were scanning Earth and, on three successive days, bounced a beam off Pennsylvania, then Kansas, then California.)

The microwave echoes on June 3 and June 5 were feeble, about what you would expect if they had struck a body of liquid. However, on June 4, the microwave echo was much stronger and was of a type that we would expect to get back from Venus. That makes it look as if the June 4 beam had struck a solid surface.

It would seem, then, that Titan may be the only body in the solar system that we know of, so far, that resembles Earth in having a surface that is partly liquid and partly solid. Titan, like Earth, may have continents and oceans, though Titan's would be altogether different in their chemical makeup from ours.

Perhaps in the future, microwaves will bring us enough information to allow us to make a map of Titan's surface and to identify the materials composing it. Is the ocean hydrocarbon or nitrogen? Are the continents rock, ice, or solid carbon dioxide? Scientists would like to know, and they may yet find out.

Space Rocks

Meteorites are rather rare objects, but they have always been important. Now, however, we have a new place to look for them and a new technique being developed to aid in the search.

When human beings first began to make use of metal, copper and bronze were the best they could find. However, they occasionally came across lumps of metal in the ground that could be forged into spearpoints and plowshares that were harder and tougher and could hold an edge sharper and longer than anything made of bronze. The ancients did not know it, of course, but these were nickel-iron meteorites that had fallen to Earth from outer space.

On very rare occasions, meteorites were seen to fall and hit the ground. The awed onlookers might naturally consider them products of heaven sent to the Earth by the gods, and as a consequence, they might have worshiped them. In Mecca, the Black Stone (supposedly given to Abraham by the angel Gabriel) in the Kaaba, Islam's holiest shrine, is probably a meteorite.

So earnest was the search for these strange and useful lumps of metal that there are no iron meteorites to be found in the Middle East, where civilization first developed. They were all located and made use of long ago. It was not till 1500 B.C. that people in the Middle East learned to smelt iron from its ores and were no longer dependent on the occasional meteorite for the metal.

In modern times, meteorites have been put to more scientific use. They are old indeed, dating back to the beginning of the formation of the solar system. The Earth itself is also ancient, of course, but the Earth has undergone such geological disturbance in the course of its long history that the oldest unchanged rocks found in Earth's crust are only a little over 3 billion years old. The smaller the object, the less disturbance it undergoes, and meteorites are so small that they have undergone virtually no disturbance at all.

The close study of radioactive changes in meteorites has caused scientists to conclude that the Earth, and the entire solar system with it, is about 4.6 billion years old.

Meteorites exist in different varieties. Those of nickel-iron composition are easily recognized because, except in meteorites, lumps of metal of this sort do not occur in Earth's crust. These make up only 10 percent of all meteorites, however. The rest are almost always rocky in nature, and unless they are actually seen to fall, they usually go unnoticed among Earth's own rocks, unless they happen to be picked up and studied for other reasons.

A very few meteorites are *carbonaceous chondrites*. These contain a certain amount of water tied to the rock molecules that make up the main body of the meteorites. They also contain carbon-containing *organic* molecules, including fats, amino acids, and so on. These organic molecules are similar to those found in living creatures on Earth, but they are not the product of life. Some unmistakable characteristics of the organic molecules in meteorites show them to be formed by natural nonlife processes.

This hints that when Earth was formed, such organic molecules may have formed very early and may have begun to march toward greater complexity and life. In short, the study of such meteorites may give us hints to the origin of life.

Scientists are very eager to study as many meteorites as possible, for such information as they might yield that can't be obtained in any other way (so far). However, nickel-iron meteorites are rare and carbonaceous chondrites are even rarer. And, unfortunately, the most common "stony" meteorites usually are undetected against the rocky background of Earth's land areas.

There is, however, one land area on Earth that is not rocky, at least on the surface. This is Antarctica, which consists of 5 million square miles covered by a thick layer of ice. Against this covering of ice, any occasional rock would be conspicuous indeed, and such rocks are almost certain to be meteorites. In the last few years, a number of meteorites have been located and plucked off the Antarctic ice sheet for study. The largest found so far is about two feet across and weighs 240 pounds.

What makes them particularly useful is that such meteorites, located in Antarctica's sterile ice, are less likely to have been invaded and changed by microscopic life than are those that fall in more salubrious climes.

However, until now, only meteorites brought to the surface of the ice by slow glacial movement have been discovered. There must be many more meteorites buried beneath the surface. Scientists at the Naval Air Development Center in Warminster, Pennsylvania, have run experiments that show that such buried meteorites can be located by radar, even if they are buried several dozen feet beneath the ice and weigh no more than a pound or two. The entire Antarctic continent can be surveyed in this way in the future, and an incredible bonanza of meteoric material may be located and, eventually, dug for.

Close Call with an Asteroid

On March 23, 1989, Earth had another close call. A small asteroid, about half a mile across, whizzed past at a distance of about 500,000 miles, about twice the distance from Earth to the Moon. This sounds like a safe distance, of course, and "a miss is as good as a mile," they say.

However, this rock is following an orbit that nearly crosses Earth's and every once in a while (a fairly long while), both Earth and asteroid reach the crossing point at the same time and it gives us one of those close-call whizz-bys.

We might argue that 500,000 miles, or possibly a little closer, may be as close as it can possibly come if the orbits stay as they are, but they won't. Earth is a massive body and its orbit is quite stable, but the asteroid is a tiny thing in comparison and it is subject to the pull of Earth and Moon and Mars and Venus as it moves so that its orbit is constantly shifting a little.

Its orbit may shift so as to carry it farther from Earth or possibly closer, so the chance of its moving into a collision course is very small, but it's not zero.

The trouble is that this asteroid is not the only one. Back in 1937, an asteroid that astronomers named Hermes passed within 200,000 miles of Earth, and it was larger than the one that just missed us. It may have been a mile across.

And on August 10, 1972, a small object, only forty feet across, perhaps, actually passed Earth at a distance of only thirty miles above the surface of southern Montana—and whizzed on. It had passed through our stratosphere.

Some astronomers think there may be at least a hundred objects that are half a mile or more across and that have orbits that send them skimming by Earth. And there may be thousands that are a few dozen feet across. Obviously the chance that some one out of all these will finally strike Earth is far greater than the chance that some specific object, such as the one that just missed us, will do so.

Even a comparatively small object like the one that missed southern Montana could do fearful damage if it struck. If it hit land, it would gouge a sizable crater. After all, these projectiles can be moving twenty miles a second as they hit Earth.

An object half a mile across, like the one that passed us in March, would hit with the force of 20 billion tons of TNT. If it struck New York, it would undoubtedly wipe out the entire city and kill millions of people in an instant. If it struck the ocean it might be even worse, for the water in the ocean would slosh tremendously and huge *tidal waves*, mountains of water hundreds of feet high, would crash onto neighboring shores and drown tens of millions.

If the object is bigger still, it might puncture Earth's crust, set off volcanic action, start worldwide forest fires, drown half the continents, and cast so much dust up into the stratosphere as to cut off sunlight for a long period. Such a huge strike might kill most or all of life, and, indeed, such a huge strike is thought by many to have wiped out the dinosaurs 65 million years ago.

Lesser collisions have happened recently. In Arizona, there is a crater four-fifths of a mile across and six hundred feet deep that formed perhaps 50,000 years ago as a result of a collision. It probably didn't kill anyone because human beings had not reached the Americas yet. In 1908, a much smaller strike in central Siberia knocked down every tree for twenty miles around but in a desolate, uninhabited region, and nobody was killed.

In fact, there is no record in historic times of any human being's being killed by a meteoric strike, but we can't stay that lucky forever.

What can we do about it?

Thirty years ago I wrote an essay that appeared in the August 1959 issue of *Space Age* called "Big Game Hunting in Space." In it, I advocated the establishment (once we had the capability) of a space sentinel that would watch for any nearby object more than a few feet wide that might be approaching. It might then be blown up by a hydrogen bomb planted in its path or by something more advanced. (It would be a "Star Wars" defense aimed at asteroids rather than at enemy missiles.)

As far as I know, I was the first to suggest this, but since then astronomers have discussed the problem very seriously. After all, there are estimates that a "city-busting" strike might happen, on the average, once in 50,000 years, and it's been that long since the Arizona crater. It may be due, so to speak.

Of course, if we smash up a small asteroid, the rubble will continue in its orbit, perhaps, but then if this does strike the Earth, each individual bit can't do much damage. Instead of having a huge crater gouged out, we'll be treated to a brilliant meteor display as the small pieces burn up in the air or land as small rocks.

Diamonds from Space

Scientists find small quantities of various substances in meteorites and no longer expect surprises. However, not too long ago a group of chemists working under Edward Anders of the University of Chicago were enormously surprised to find diamonds in meteorites.

This doesn't mean they were suddenly wealthy, of course, for the diamonds they found are microscopic in size. There were two types of diamonds, one so small that if you placed about 250,000 of them side by side they would form a line an inch long. These were the larger ones. The other type was so small that it would take 10 million of them side by side to stretch across an inch. The chemists were delighted, however. Meteoric diamonds, even that tiny, represent wealth of another kind: wealth of knowledge.

The solar system, including the Sun and all the planets, condensed out of a vast cloud of dust and gas aeons ago. In the process, most of the material grew very hot and underwent considerable change. It is hard to tell from the makeup of the Sun or of the Earth just what the original cloud was like.

Small bodies, however, underwent fewer changes than large bodies. Thus the small meteors that fly about in interplanetary space can tell us more about the beginnings of the solar system than anything larger in the solar system can. It was through studies of meteors, in fact, that we learned the age of the solar system: 4.6 billion years.

But even the dust cloud out of which the solar system

was born had a history of its own. It didn't exist unchanged throughout the lifetime of the universe. Originally, the cloud consisted entirely of hydrogen and helium, the two simplest atoms. However, stars form more complicated atoms and send some of them off into space in *stellar winds*. (Our own Sun has a *solar wind*.) Red giant stars, which are huge and rather unstable, are much more active in this respect. As a result, interstellar gas clouds are contaminated with heavier atoms. Sometimes, stars explode as supernovas and then vast quantities of complex atoms are hurled into space and gas clouds grow even more contaminated.

The cloud out of which the solar system formed was heavily contaminated in this way, for our Earth and our own bodies consist very largely of complex atoms that had their origin not in the dust cloud but in the contaminating stars. (As astronomers sometimes say, we are "star stuff.")

As the material in the dust cloud condensed to form the solar system, so much change took place, even in forming small bodies such as meteorites, that we can't tell much about the nature of the contamination of this cloud. However, one substance, and only one, was tough enough to withstand these changes and give us some clues to the details of the contamination. That substance is diamond.

One of the elements formed copiously inside stars is carbon. Carbon atoms usually clump together rather loosely in the form of graphite. Apparently, stellar winds and stellar explosions cause some graphite to clump together tightly into diamonds, the hardest substance known.

The tiny diamonds in meteorites are not pure carbon, however. Inside their structure are still tinier bubbles containing gases. These gases appear to date back to the original cloud, shielded from change over the millennia by the enclosing diamond shell.

In red giants, the formation of the more complex atoms takes place slowly, by the addition of tiny particles called "neutrons," one by one. This means that the atoms that are finally formed tend to contain relatively few neutrons. On the other hand, when a star explodes, all the atomic changes take place rapidly. Neutrons are forced into atoms at an enormous rate, so that the final atoms that are formed tend to contain relatively many neutrons.

It turns out that the two types of diamonds in meteorites had different origins. Both types of diamonds contain tiny bubbles of the rare gas xenon, but in the case of the larger diamonds the xenon is predominantly of the type known as xenon 130. Each atom of this type contains seventy-six neutrons. In the case of the smaller diamonds, the xenon contained is predominantly of a type known as xenon 136, each atom of which contains eighty-two neutrons.

It would seem, then, that the larger diamonds arise from the stellar winds of red giants, and the smaller diamonds arise from exploding supernovas.

This tells us something about the nature of the contamination of the original cloud right away, and there seems to be no doubt that further studies will yield further information. It may be important to learn why stars form diamonds at all, instead of the more easily formed graphite. After all, as much as one one-thousandth of all the carbon in space may be in the form of diamonds. Why should that be?

The Dead World

The stage is set for a new race to the Moon.

The Soviet Union is building a shuttle fleet of its own and already has a simple space station. The United States is refurbishing its own shuttle fleet in light of the lessons of the *Challenger* disaster and is planning quite an elaborate space station.

The United States won the first race to the Moon, but this was just a tour de force. We paid the Moon a series of short visits and then retired. The new race will be for a far greater prize: It will be for the establishment of a permanent base on the Moon.

But why? The Moon is an entirely dead world, just an uninteresting hunk of rock. Why bother with it?

Actually, the Moon is an enormous piece of real estate right in our backyard—only three days away. Its surface is equal in area to that of the two American continents put together, and the fact that it exists at all is amazing. There are six other large satellites in the solar system, and all of them belong to giant planets. That a small planet like Earth should have such a large satellite as the Moon is still unexplained.

And it's a good thing that the Moon is a dead world. If it had life, even the simplest form of life, we might feel compelled to leave it untouched: to study its life and to protect it as we try to protect the California condor. The Moon would belong to its life. As it is, the Moon belongs to no one, not even to the simplest virus. Human beings can make use of its resources freely.

And it has resources. Moon soil is rich in metal-containing minerals. The lunar soil can be smelted and made to produce all the structural metals: iron, aluminum, titanium, magnesium, and so on. It can be made to form cement, concrete, glass. It can even be considered a rich source of oxygen. All these materials can be used to build structures in space, indefinite numbers of them.

You might wonder why we have to go to the Moon to get the material. Don't we have plenty of all those things in Earth's own soil? This is true, but the resources of Earth belong to the people of Earth, who have great need of them and might not welcome having countless millions of tons of metal and other materials diverted for structures in space. The resources of the Moon are another matter. They have been lying there unused for billions of years. If we use them now, we deprive nobody.

There is another reason for using the Moon's resources rather than Earth's. The Moon is a smaller world and has only one-sixth the gravity that Earth has. To lift a ton of material off the surface of the Moon and hurl it into space takes only a small fraction of the energy needed to lift that same ton off the surface of the Earth.

What structures can we build? To begin with, we might build solar power stations in space that could collect sunlight with up to sixty times the efficiency of similar stations on Earth's surface. This solar energy could be beamed down to Earth as microwaves and solve our energy problems forever.

We might build automated factories that could take advantage of the unusual properties of space (extreme vacuum, zero gravity, energetic solar radiation, and so on) to manufacture devices and carry through processes in ways that could not be duplicated on Earth.

We might build observatories to investigate the universe in ways we can't on Earth's surface, where the atmosphere continually obscures things. We might build laboratories to make studies that can't be conducted on Earth's surface or to conduct biological experiments that might be too dangerous to conduct on Earth.

We might even build artificial towns in space, each capable of holding ten thousand (or more) men, women, and children.

Making use of lunar resources (plus a little from Earth, because the Moon is lacking in the important elements carbon, hydrogen, and nitrogen) we can build the beginnings of a space-centered society and lay the groundwork for its expansion to the asteroid belt and beyond in the forthcoming century or two.

It seems quite certain that the Soviets, who have been pushing steadily ahead in space, are looking forward to something like this. And we must, too. The possible gains to be had from this widening of the human range, both physical and psychological, are enormously greater than the money, effort, and risks involved, and we don't want the Soviets to reap them alone.

It seems to me, of course, that we don't want to adopt a confrontational attitude, either. Both we and the Soviets can advance faster if we cooperate. In fact, so huge is the task of establishing civilization in space that we should consider it a global project. Both we and the Soviets should welcome not only each other's help but whatever help can be given by every other nation on Earth.

The Slow Breakdown

If we can ever establish a permanent base on the Moon, what amazing things we might be able to do as a result! For instance, we might be able to find out whether certain basic theories of nature are correct.

Physicists have, in recent years, evolved what are called the "Grand Unified Theories," in which the forces of nature are gathered into one set of mathematical relationships. Such a theory, if it proves valid, might finally tell us how the universe began, how it developed into its present state, and what its ultimate fate might be.

But how do we know whether the Grand Unified Theories are correct? One way is to see whether they suggest any previously unsuspected phenomena and then run experiments to test these hypotheses.

For instance, ever since the proton was discovered about three-quarters of a century ago, it has seemed to be a stable particle. Left to itself it would, apparently, last through all eternity.

According to the Grand Unified Theories, however, the proton should have a small, an incredibly small, tendency to break down. In fact, in something like 200 million trillion trillion years, half of all the protons in the universe should have broken down. This is an extremely long time, to be sure. It is about 13 billion trillion times as long as the entire lifetime of the universe so far. This means that in the time since the universe came into existence, only a very tiny fraction of the protons have had a chance to break down.

How, then, can we check the correctness of the Grand Unified Theories to find out whether, in actual fact, protons are breaking down at a very slow rate? Obviously we can't wait trillions and trillions of years to check the matter.

Well, we don't have to. Even if it takes a virtual eternity for many protons to break down, a very few are constantly breaking down all about us. For instance, twenty thousand tons of water, or of iron, would contain millions of trillions of trillions of protons, and about twelve of these would break down in the course of a year. This is an insignificant percentage, but each proton, as it broke down, would produce particles that could be detected, and if those twelve per year were detected, they would constitute a strong piece of evidence in favor of the Grand Unified Theories. After all, if these theories were not true, there would be no breakdowns at all.

Sensitive devices have been set up to detect such very occasional proton breakdowns, and, so far, they have yielded nothing. Perhaps this means the Grand Unified Theories are not true, but scientists are not ready to admit this yet. For one thing, there is a feeling that the detection devices are not yet quite sensitive enough to do the job. For another, even if they were, there is interference.

There are, after all, various kinds of energetic radiation all about us, from sunlight to cosmic rays. These produce particles that represent "noise" in the detectors and mask true proton breakdowns.

To cut down on the noise, such detectors are placed far underground. This produces a "quiet" background, with one exception. Cosmic rays that constantly bombard the Earth react with atoms in the Earth's atmosphere to produce tiny particles called *neutrinos*. These neutrinos hardly interact with matter and pass through the entire Earth as if it were empty space. They can pass through the detectors, however deep underground they are.

Very occasionally, such neutrinos interact with protons to produce particles similar to those that should be produced by proton breakdown. For every true proton breakdown caught by the detectors, nearly 100 neutrino interactions would also be caught. Disentangling the proton breakdowns would be a difficult task.

Ah, but what if we were on the Moon, where there is no atmosphere? In that case, we might burrow a tunnel 330 yards long, 16 yards wide, and 8 yards high into the side of a crater about 100 yards below the surface. A set of very massive and complicated detectors would be placed here and there in the sides of the tunnel.

Cosmic rays hit the Moon, too, but with no atmosphere present, the number of neutrinos formed would be much lower. Scientists calculate that there would, under such conditions, be only one neutrino interaction for every two true proton breakdowns. If, then, we could set up this complicated and very expensive experiment on the Moon, the supreme quietness of the background would allow us to check the validity of the Grand Unified Theories with comparative ease.

Old Reliable

There are some things that we would want very much to be reliable—the Sun, for instance. We don't want it to get noticeably bigger or smaller or hotter or cooler. It's just right the way it is, thank you, and a recent bit of research indicates that it is, indeed, staying just right.

We're pretty sure that it's been reasonably reliable through all of Earth's history. If the Sun ever got hot enough to boil Earth's ocean, or cold enough to freeze it nearly solid, any life present would surely be destroyed. But, as nearly as we can tell, life has been continuous on Earth for at least 3.5 billion years.

There have been abnormalities, of course. In the last million years there have been a number of ice ages, and every few tens of millions of years, there is a great wave of extinctions. As nearly as we can tell, however, the Sun is not directly involved in these disasters. Instead, they come about through meteoric impacts or through changes in distribution of the continents or depth of the ocean. At least, so we think.

But even if the Sun is reliable in the long run, might it be that just now it is entering a period of slight unreliability? Might it perhaps be undergoing changes that are not large enough to endanger life in general but large enough to be very inconvenient to human beings?

For instance, there has been some talk in recent years that the Sun has been shrinking slightly during the last few centuries.

Right now, the Sun is 1,919 arc seconds in diameter, but some astronomers have found reasons for supposing it might have been 1,927 arc seconds in diameter back in 1700. That's not much of a shrinkage, but it could be a signal of potential problems ahead.

Is there any way of checking?

There may be. Every once in a while the Moon passes directly in front of the Sun, and the Moon's shadow is cast on the Earth. The Moon's shadow narrows as it approaches the Earth, and by the time it reaches the Earth's surface, it is, at most, 170 miles across. The exact width of the shadow depends on the distance of the Sun and Moon from the Earth on the day of the eclipse, together with the diameter of the Sun and Moon. The distance of the Sun and Moon and the diameter of the Moon have certainly not changed measurably in the last few centuries, so this leaves only the diameter of the Sun as uncertain.

If the Sun had been wider three centuries ago than it is now, its light would have overlapped the Moon a trifle more and narrowed the shadow to less than it would be nowadays. So all we have to do is measure the width of the shadow of an eclipse that took place three centuries ago. But how do we do that?

Here we have a lucky break. There was an eclipse on March 3, 1715, when the Moon's shadow cut across southeastern England, which was well advanced scientifically at the time. What's more, there was in England at the time one of the greatest astronomers of the age, Edmund Halley. (He was the same astronomer who was the first to work out the orbit of Halley's comet.)

Halley organized the viewing of the 1715 eclipse by amateur observers all over England, and he collected all the eyewitness accounts. Each account, for instance, described exactly how long the eclipse lasted. The deeper into the shadow, the longer it would last (a little over seven minutes is the maximum for any eclipse). Near the edge of the shadow it would last only a few seconds.

British astronomers, headed by Leslie V. Morrison, have now gone over those accounts and recently reported the results. They found that at the southeastern tip of England, there was an account from one Will Tempest, who lived near Cranbrook, Kent. He reported that the eclipse lasted only an instant. He must have been almost precisely at the southern edge of the Moon's shadow.

There was also a report from a Theophilus Shelton, who lived near Darrington in West Yorkshire, and he too said that the eclipse had only lasted for an instant. Indeed there was one part of the Sun still visible, but it was only star-size at the time. He must have been almost precisely at the northern edge of the Moon's shadow.

This had been reported before, actually, but the Morrison team managed to locate the exact position of the houses of Tempest and Shelton, instead of just going by the location of the town centers. When they did so, they were able to measure what the width of the Moon's shadow must have been to an accuracy of less than a mile, and they found that it was just what it would have been if the Sun's diameter had been exactly what it is today.

If, on the other hand, the Sun's diameter had been 8 arc seconds wider, the shadow would have been three and one-quarter miles farther south in Yorkshire and three and one-quarter miles farther north in Kent. Neither Tempest nor Shelton would have been able to see the eclipse reach totality. Enough of the Sun would have remained exposed throughout to destroy the eclipse effect. So the Sun remains Old Reliable after all.

Going Where the Energy Is

We're going to need energy for the indefinite future, and Soviet scientists are making plans with this in mind. They are now proceeding with a project to place huge solar power stations in space to beam electricity to Earth. Although the project has potential military application, it also offers an excellent opportunity for multinational cooperation that could promote world peace.

The Sun is an obvious energy source, one that will last for billions of years. Sunlight reaching the Earth can be converted into electricity, but the atmosphere absorbs some of the light and scatters the rest, dust dims it further, clouds block it still further, and night ends it altogether for part of each day.

So, why not go where the energy is? Why not move out into space? Beyond the atmosphere, the light of the Sun shines in steady splendor, with no clouds, dust, or air of any kind to diminish it.

Suppose we imagine a device that is far above Earth's equator, one that is capable of absorbing sunlight and turning it into electricity. It would move into Earth's shadow for an occasional few hours about the time of each equinox. Except for that, the device would be exposed to full sunlight continuously. It is estimated that such a device would convert up to sixty times as much sunlight into energy as the same device could on the surface of the Earth.

If such a device is 22,300 miles above Earth's equator, it will orbit the Earth in just twenty-four hours. Viewed from a spot on the equator directly under it, it will seem to remain stationary in space. The device can collect sunlight and convert it into electricity, which can be, in turn, converted into microwaves. The microwaves can be beamed down to a receiving station on Earth that could again convert it to electricity for distribution where needed.

To get a reasonable quantity of energy, however, such a device would have to pick up a great deal of sunlight. That means spreading out *solar cells* (units that convert sunlight to electricity) over a vast area. The usual estimate is that the total device will have an area equal to that of Manhattan Island or more. What's more, there would be perhaps sixty such devices placed in orbit above the equator, taking up a total area greater than that of the state of Rhode Island.

The amount of energy thus made available continuously, year after year, century after century, might be equal to the output of six hundred nuclear power plants. With time, the quantity of energy obtained would surely be further increased as the efficiency of the device was enhanced.

There are enormous difficulties in the way, of course. It could easily take fifty years of concentrated work and cost up to $3 trillion to build the necessary devices. Once built, the solar power stations in space would have to be continually maintained and repaired. In space, the devices would be safe from weather, depredations of wildlife, and human vandalism but would be vulnerable to damage from space junk. Some of this would be natural, because space is fairly full of particles of dust and gravel. Some would be human-made—bits and pieces of satellites and probes.

Then, too, there is a question as to how the microwave beams stretching from the solar devices down to Earth's surface might affect the ozone layer, the atmosphere, people, wildlife, and so on.

The project had been advanced by Peter E. Glaser of Arthur D. Little Inc. in Cambridge, Massachusetts, in the 1960s. NASA had considered the possibility in the 1970s, but cost and environmental considerations seemed to kill American interest.

Now, however, the Soviets have picked up the idea. They have a new rocket, *Energia*, which is at least four times as powerful as the best American efforts, and they hope to use it to lift into space the quantities of material that would be required for such an enormous project.

They might well start with something simple, building a passive reflector of sunlight. This would create a small "moon" in the sky that could light up cities or warm farming areas in case of unseasonable frosts. This might be done, in an experimental way, by the 1990s.

If the Soviets make progress in this notable plan, the United States might well fear possible military applications of solar power and reflectors in space. I have always felt that the surest way of countering such a possibility is to internationalize such huge space projects that may, in any case, be too expensive for anything less than a global effort.

Besides, the peaceful uses of such solar energy power stations in space should not be confined to a single nation. It makes sense to suppose that sunlight belongs to all of Earth. The common desire to make use of this energy and to maintain and improve the stations should give the nations of the world a vital project to rally round and impose a greater feeling of common purpose than now exists.

Because any serious disagreements or controversies would adversely affect the smooth workings of the stations, cutting down the energy supply for all, this might also be a strong incentive toward peace.

An Ocean of Gasoline

Imagine finding an ocean of gasoline! Actually, there may be such a thing. In fact, it may exist in two different places. But not on Earth, of course.

There are seven large satellites in our solar system, and our Moon is one of them. It is too small (2,160 miles across), with too weak a gravitational pull, to hold an atmosphere. This is true especially because the ability to hold an atmosphere lessens as temperature increases. The Moon is the nearest of the large satellites to the Sun, and its temperature is sometimes above the boiling point of water.

Jupiter has four large satellites, which get only one-twenty-seventh the heat from the Sun that our Moon gets. What's more, two of them, Ganymede and Callisto, are very large: more than three thousand miles across. Still, they are too small and too warm to have atmospheres.

Saturn has one large satellite, Titan, which is also more than three thousand miles across and receives only one-ninetieth the heat from the Sun that the Moon receives. It is large enough and cold enough to have an atmosphere. Back in 1948, G. P. Kuiper detected this atmosphere and found it to contain methane, a compound of carbon and hydrogen. Methane is the chief component of what, on Earth, we call "natural gas."

But in 1981, when the probe *Voyager 2* passed Saturn, it showed that Titan had an unexpectedly thick atmosphere, one that might be even thicker than the Earth's. The methane was there, and, in addition, a large quantity of nitrogen. (Cold nitrogen is almost impossible to detect at a great distance.)

Titan's atmosphere is hazy, so that we can't see through it to the surface. However, astronomers are familiar with the behavior of nitrogen and methane and can guess what might happen. Nitrogen, an inert gas, would not change. Methane, however, can have its molecule pulled apart by the energy of sunshine and the parts could come together into larger molecules made up of carbon and hydrogen. Methane has only one carbon atom in its molecule, but the radiation from the Sun might combine it into molecules with two carbon atoms, or three or more.

Nitrogen and methane are gases at Titan's temperature, but the more complicated molecules would be liquids. It is possible, then, that under Titan's thick atmosphere, there may be puddles, lakes, rivers, even oceans of molecules with two carbon atoms (called *ethane*), or three (*propane*), or more. Molecules with seven or eight carbon atoms would be gasoline. They might be solid at Titan's temperatures, but such solids would dissolve in the ethane-propane ocean.

What it amounts to, then, is that Titan, under the haze of its atmosphere, may possess a gasoline ocean.

If we look beyond Titan, one large satellite remains. Named Triton, it circles Neptune, the most distant of the large planets. Having passed Uranus, the planet that lies beyond Saturn and that has no large satellites, *Voyager 2* sped toward Neptune and passed it in August 1989.

Triton turned out to be considerably smaller than Titan. It is even colder, since it gets only one-nine-hundredth the heat from the Sun that our Moon does and only one-tenth the heat that Titan gets. Triton should, therefore, also have an atmosphere, but a thinner one.

However, these two worlds are not just ripe for the taking. Titan is about 886 million miles from us, and Triton is three times farther still, 2.8 billion miles away. At such distances, any gasoline we could retrieve from Titan would be expensive indeed. Besides, it wouldn't be a good idea to bring back this distant gasoline and burn it here. It would just use up our oxygen and replace it with carbon dioxide, trouble already caused from burning our own gasoline and coal.

However, the time may someday come when human beings have large settlements in the outer solar system. In this case, Titan might prove to be a valuable resource. To be sure, the gasoline it would supply would undoubtedly not be needed for energy purposes because the distant settlements should, presumably, be running on nuclear fusion reactors.

However, these distant worlds would contain surface materials made up of nitrogen, carbon, and hydrogen, three elements that are essential to maintain the settlements. These elements are relatively rare on most of the worlds we can approach. (The Moon has none of them, for instance, so lunar colonists will depend on Earth for their supply.) The distant outposts might therefore be thankful to get what they need from Titan and Triton.

The Elusive Tenth

For a century astronomers have been searching for a large planet beyond Neptune—and they have not found one. Now, however, they have a new tool for the purpose that may help. It is a probe that is still broadcasting though it is far beyond the outermost known planet.

What makes astronomers think there is such a planet in the first place?

After Uranus, the seventh planet, was discovered in 1781, its orbit turned out to be slightly off the calculations. Astronomers decided that there must be an eighth planet, beyond Uranus, whose gravitational pull on Uranus wasn't being allowed for. In the early 1840s, astronomers began to calculate where the eighth planet would have to be to account for Uranus's actual motion. In 1846, the indicated spot in the heavens was examined, and a new planet, Neptune, was discovered after a mere half-hour of searching.

In 1900, therefore, astronomers began to calculate the possible position of a large planet beyond Neptune. This time the search was far more difficult. The farther off a planet, the dimmer it is and the harder it is to see against a background of dim stars. What is worse, the farther a planet is, the more slowly it moves, making it even more difficult to distinguish from the motionless stars.

In 1930, however, a ninth planet, Pluto, was discovered. It existed beyond Neptune, and, for a while, it seemed the problem had been solved. However, the longer Pluto was studied, the smaller it seemed to be. We now know that it is smaller than our Moon and scarcely larger than a large asteroid. It is far too small to have any noticeable gravitational effect on Uranus or Neptune.

This means that astronomers are still searching for some large object beyond Neptune, something that, when it is found, would qualify as a tenth planet. No such planet has been sighted so far.

Back in 1972, however, a Jupiter probe called *Pioneer 10* was launched and later a sister probe, *Pioneer 11*. In 1973 and 1974, the probes passed Jupiter and have continued to move away from the Sun ever since. By now *Pioneer 10* is far beyond the orbit of Neptune. Pluto is now slightly closer to the Sun than Neptune. *Pioneer 10*, therefore, is about a billion miles farther from the Sun than any known planet.

Pioneer 10 is continuing to emit radio waves of a very precise wavelength. This wavelength changes slightly with the probe's speed of motion. Astronomers can calculate exactly how the speed and the wavelength change because of the gravitational pull of the Sun and of the various known planets.

If there is any change in the radio wavelength beyond that which is calculated, it must be the result of a gravitational pull. There are three possible sources of such a pull. One is the belt of distant comets that is supposed to exist far beyond the planetary orbits. This is not a very likely source because the comets lie in every direction and the gravitational pulls tend to cancel. A second source is a possible dwarf star that might be a distant companion of the Sun. Finally, a third possibility (and the most likely) is that elusive tenth planet.

In recent years, however, nothing that has been detected from *Pioneer 10* would indicate the presence of an unsuspected gravitational field. This is taken to mean that there is not likely to be a companion star to the Sun or even a really large planet out there close to the size of Jupiter, for instance. (Jupiter has over three hundred times the mass of the Earth.)

Still, a moderately large planet, say five times the mass of the Earth, might exist. It might be showing no effect on *Pioneer 10* because it could be in a part of its orbit where it would be too distant to have a noticeable gravitational effect. (A companion star or a Jupiter-size planet would show an effect at any point in any reasonable orbit, but a smaller planet would not.)

The tenth planet, therefore, may have a very elliptical orbit that brings it fairly close to the outer planets only every eight hundred years or so. It would then stay close enough to have a gravitational effect for about one hundred years and diminish to virtually nothing for seven hundred years.

Perhaps, then, the tenth planet was close enough to alter the orbits of Uranus and Neptune, very slightly, between 1810 and 1910, but not since then. There would not be another period of interference until A.D. 2500 or so. What's more, the planet may be moving in an orbit at a wide angle from that of other planetary orbits. This would make it present in totally unexpected parts of the sky and make it that much harder to find. Therefore, astronomers must continue to hunt.

The Little Rocket that Could

A rocket probe that was launched from Earth on March 2, 1972, is still moving outward and sending useful messages back to us nearly eighteen years later.

The rocket is *Pioneer 10*, whose original mission was to explore Jupiter and its neighborhood. It passed Jupiter on December 2, 1973, twenty-one months after launch, and gave humanity its first closeup look at that giant planet. Speeded by Jupiter's gravitational field, it then moved on into the outer solar system and by mid-June 1983 it had passed Neptune's orbit. Because Pluto was at that time no farther than Neptune was, *Pioneer 10* had traveled beyond the planetary system.

Years more have passed and it is still traveling on its long voyage, well over 4.2 billion miles from the Sun. The Earth itself is only 93 million miles from the Sun, a distance that is called an *astronomical unit* (AU). This means that *Pioneer 10* is now about 45 AU from the Sun. The distant planet Pluto, at the far point of its orbit, is 47 AU from the Sun. But it is at the near point now and won't be at the far point for another century. From this mighty distance, radio waves are still being sent out by *Pioneer 10* and are still being received on Earth. Traveling at the speed of light, it takes those waves six and a quarter hours to reach us.

But what can *Pioneer 10* find to tell us about the vast emptiness beyond the planets?

Here is one thing: the Sun is hot and active and sprays charged particles, chiefly protons and electrons, into space.

These travel outward in all directions at high speeds. This phenomenon was first actually detected in 1962 by the rocket probe *Mariner 2* as it approached Venus. The speeding particles are called the *solar wind.*

The solar wind is important. It reaches Earth, produces the auroras in the polar regions, and fills Earth's magnetic field with charged particles. Occasionally, there is a vast explosion on the Sun called a *solar flare,* and then the solar wind rages intensely for a while, producing magnetic storms on Earth and upsetting electronic communications. The intense concentrations of charged particles can threaten the lives of astronauts, so the matter of the solar wind will become ever more important to us as we move out into space.

As the solar wind moves outward, it spreads over a larger volume and becomes thinner. Eventually, it must thin out and fade into the faint wisps of gas that fill outer space. Before *Pioneer 10* went on its mission, scientists thought that the solar wind would fade out not far beyond the orbit of Mars.

However, *Pioneer 10,* 45 AU from the Sun, is still detecting a perceptible solar wind even though it is far beyond the orbit of Neptune. Now scientists think the solar wind must reach as far as 50 to 100 AU from the Sun before fading into the general interstellar background. *Pioneer 10* ought to continue sending messages for as long as ten more years, and it may reach the end of the solar wind before it dies.

Here is something else: Einstein, in his theory of general relativity, predicted that every object that is moving in orbit gives off "gravitational waves" and loses energy in that fashion. The waves are so weak, however, that the loss of energy is infinitesimal.

Scientists are eager to detect these gravitational waves. In the first place, to do so would be another piece of evidence in favor of Einstein's theories. In the second place, violent events involving huge masses—such as the collapse or collision of stars or the activities of black holes—should liberate intense gravitational waves that might yield information about such events that nothing else could.

Unfortunately, even the most intense gravitational waves are so weak that we don't have instruments delicate enough to detect them. Scientists have set up huge aluminum cylinders that quiver when gravitational waves pass over them. The quiver is less than the diameter of a proton, however, and it is difficult to detect it from among all the possible phenomena that might make such cylinders quiver.

Pioneer 10, however, is far out in space, in a kind of ultimate "silence," where there is nothing to produce vibrations as the rocket probe coasts silently through emptiness. All that can reach it are gravitational forces that express themselves in these feeble, feeble waves. By early 1989, equipment on board *Pioneer 10* was prepared to try to detect these waves. If the experiment succeeds, the last major prediction of Einstein's general theory of relativity will be confirmed and *Pioneer 10* will once again be the little rocket that could.

The Lopsided Satellite

Voyager 2, which made a successful fly-by at Uranus in January 1986, has been racing toward Neptune, the fourth and farthest of the giant planets, and was on a course to reach it in August 1989. In doing so, it will study not only that planet, but also its two satellites.

Neptune itself will probably turn out to be very much like Uranus. Of course, being farther from the Sun, it is colder. Of Neptune's two satellites, one, named Triton, is about twenty-four hundred miles across, which makes it a little larger than our Moon. It is about as far from Neptune as our Moon is from the Earth. It is very likely that Triton will closely resemble Titan, the largest satellite of Saturn, except that Triton will also be colder. Triton, like Titan, may well turn out to have a thick atmosphere of nitrogen and methane, and there may be lakes and seas of liquid nitrogen on Triton's surface.

Neptune, however, has a smaller satellite, Nereid, which may turn out to be the exciting news of the fly-by, for it is a strange world indeed.

Nereid is so far from us and so small that it wasn't discovered until 1949, a full century after Neptune and Triton were discovered. It is hard to tell just how large it is, but two astronomers at the Goddard Space Center in Greenbelt, Maryland, Martha W. Schaefer and her husband, Bradley E. Schaefer, have recently estimated that it is just a little more than four hundred miles across. This makes it a small satellite but not a tiny one.

Its orbit, however, is extraordinarily lopsided. At one end of that orbit it moves in to a distance of only 864,000 miles from Neptune. Then it moves steadily outward, and, at the opposite end of its orbit, it is 6 million miles from the planet.

Of all the satellites, it has the most elongated orbit. It may be that it was an asteroid that ventured too close to Neptune in ages past and was captured. Or perhaps, it was one of the small bodies that clumped together 4.5 billion years ago to form Neptune but was so far from the rest that it managed to retain its independence. If so, Nereid might tell us something about the original "planetesimals" out of which the outer planets were formed.

The Schaefers have been studying Nereid from Earth, and they find that the light that it reflects is not quite like that observed in any other asteroid or satellite. This, in itself, would seem to indicate that there is something odd about it.

Even more surprising is the fact that its light is as lopsided as its orbit. The light reflected from Nereid varies in intensity, so that it gets brighter and dimmer in a periodic way. This is not in itself unusual because some other satellites and asteroids also vary in brightness. The amount of variation in the case of Nereid, however, is quite large. The Schaefers report that it is four times brighter at some times than at others.

Usually, a periodic change in brightness means that an astronomical object is turning and, as seen from a certain angle, is brighter than when seen from other angles.

One reason for this is that the object may be irregular in shape. The asteroid Eros, for instance, which can come as close as 14 million miles to Earth, is brick-shaped. When it turns so that the narrow end of the brick is toward us, less of it reflects light to us and it seems dimmer than when its broad side is facing us.

But can this be so in the case of Nereid? An irregular object has to be small. Eros is only fifteen miles across. A large object has a stronger gravitational pull, and this pull forces its material to compact itself into a sphere. It is estimated that any object more than 250 miles across has to be spherical in shape, and Nereid is 400 miles across. Nereid, therefore, has to be a sphere and would seem the same size to us no matter how it turns.

It may be, then, that the surface isn't uniform. Some of it is light (icy, perhaps) and reflects considerable light. Some of it is dark (rocky, perhaps) and reflects little light. The planet Pluto, for instance, dims and brightens every 6.4 days. This represents its time of rotation, as it turns brighter and darker areas toward us.

Then, too, the planet Saturn has a satellite named Iapetus that shows an even greater variation in light intensity than Nereid does. Iapetus has been seen at reasonably close quarters, and it turns out that half of its surface is coated with ice and the other half with some dark substance.

However, astronomers have not yet worked out how Iapetus became a two-toned satellite. Is it that ice formed on only half the surface? Or is it that ice covered the whole surface, half of which then was covered by another, darker substance? If the latter, what is the dark substance, where did it come from, and why is it concentrated on only half the surface?

When *Voyager 2* passes within 2 million miles of Nereid, this world may turn out to be another two-toned satellite. Perhaps it will give us hints of how it got that way, hints that Iapetus did not.

[NOTE: This essay was written before *Voyager 2* reached Neptune. As it turned out, Neptune was quite different from Uranus, and Triton was quite different from Titon. No significant news concerning Nereid was delivered.—I. A.]

Beware the Flare!

There is a danger to the lives of astronauts that is always present but that they have been fortunate enough to have avoided so far. It is something called the *solar flare*. The Sun constantly shoots out charged particles in every direction. This stream of charged particles (called the *solar wind*) can be deadly if it is intense enough. Ordinarily, however, it is *not*. Every once in a while, though, a brief but violent explosion rips some portion of the Sun's surface. This is a solar flare. From the flare, a temporary but very intense stream of charged particles hurtles outward into space. If the ordinary stream is a solar wind, this is a solar hurricane.

In August 1972, for instance, there was a very strong flare, the strongest that astronomers have observed since such flares were first seen 130 years ago. A beam of strong radiation shot out; if astronauts protected by nothing more than a spacesuit had been exposed to it, it would have killed them. Fortunately, that flare took place just between the flights of *Apollo 16* and *Apollo 17*, and there were no human beings in space at the time.

This kind of good fortune won't last forever, and it would be very useful if we were to find out just when such flares took place in the past in order to see whether we can note some kind of regularity that would enable us to predict, at least roughly, when the next flare might be likely to come. We could then make sure that astronauts remained under cover during that period.

But can we look back in time to see when flares have taken place? Yes, we can. Here's how it works.

When strong radiation hits the Earth's atmosphere, some of it is bound to strike nitrogen atoms, and when it does, it sometimes converts those nitrogen atoms into what scientists call *carbon 14*. Carbon 14 is a radioactive variety of carbon that slowly breaks down until half of it is gone in 5,730 years. Although it breaks down, more is always being formed so that there is a balance, and the atmosphere always contains a tiny quantity of carbon 14.

Plants absorb carbon dioxide from the air and convert it into the molecules that make up plant tissue. Most of the carbon they use consists of ordinary stable carbon atoms, but they also pick up a tiny quantity of carbon 14. Plants, therefore, always contain some carbon 14.

Once a plant dies, however, it no longer absorbs carbon dioxide and the carbon 14 it already contains slowly breaks down without being renewed. In this way carbon 14 can be used to date dead wood. The lower the carbon 14 content, the longer it has been since the wood was part of a living plant.

It is possible to study the tree rings of trees, living and dead, and work up a calendar from the pattern, because the pattern for any given stretch of years is distinctive and unique. Such a tree-ring calendar has been traced back for nine thousand years and more.

This tree-ring calendar matches the carbon 14 calendar, for the older the wood, according to the tree-ring calendar, the lower the carbon 14 content.

Now comes the interesting part. Mostly, the carbon 14 is produced by cosmic rays and by the solar wind. Ordinarily, these are steady effects. Every once in a while there may be a supernova close enough to Earth to result in a surge of cosmic rays, and every once in a while there may be a flare that will result in a surge in the solar wind.

In either case, whether a supernova or a solar flare, there is a sudden small surge in carbon 14 in the atmosphere. The concentration doesn't remain high, for the input quickly dies down as the supernova or the solar flare subsides. The extra quantity in the atmosphere then breaks down. However, while the concentration of carbon 14 remains high, plants absorb it, producing a high concentration of carbon 14 in their tissues.

The difference between the two effects is that a supernova may only occur once in several centuries and is usually so noticeable that we know when it took place. A solar flare, however, takes place much more frequently but until quite recent years was never noticed.

If particular tree rings are carefully analyzed for carbon 14 content, one tree ring may be found to be a little high, and the year in which that high content occurred can be determined. If it was a year in which no supernova exploded, then it is clear it was a year in which a large solar flare took place.

Tree rings are particularly useful in Arizona, where the dry climate has preserved wood for a long time. Scientists under the lead of Paul E. Damon of the University of Arizona are therefore undertaking a project of analyzing tree rings for carbon 14. The result may be that a series of "flare-years" may be determined, and perhaps these may be matched with the sunspot cycle. We may know better how to protect astronauts, then.

Skimming the Sun

The worst disaster that can befall any object in the solar system is to strike the Sun. But partly because smaller, difficult-to-detect objects are more likely to meet such a fate, until recently scientists had never actually seen this happen. But such fiery ends may now have actually been observed by satellites especially designed to study the region of the Sun.

The sizable objects that come closer to the Sun than any others are comets. A number of comets have orbits that lead them into the inner solar system, skim them past the Sun, and then send them out into the vastness beyond the planets. Some approach the Sun more closely than others, and those that approach very closely indeed are called *sun-grazers*.

Astronomers, observing the heavens through their ground-based telescopes, have in the past observed eight sun-grazers, within 5 million miles of the Sun's surface or less—in some cases, much less. The most remarkable sun-grazer was a comet that skimmed the Sun's surface in 1963. At its closest approach to the Sun, it was only 60,000 miles above the Sun's surface. This is only one-fourth the distance of the Moon from the Earth's surface.

If it were conceivable that we could imagine ourselves on the comet's surface at its nearest approach to the Sun, we would see the Sun stretching two-thirds of the way from horizon to horizon and taking up half the entire area of the sky. The comet would be receiving 53,000 times as much light and heat as the Earth gets.

As it happens, comets are mostly ice. How do they endure all that heat? Why don't they just melt and vaporize instantly and vanish as a blob of expanding steam?

To begin with, comets don't stay in the vicinity of the Sun very long. The Sun's gravitational pull is stronger and stronger as the distance between the comet and the Sun grows smaller, and that means the comet is lashed along its course faster and faster. The comet of 1963 was moving at least sixty-three miles per second as it skimmed the Sun. It passed the Sun and started to retreat in a little over three hours.

Even so short a time so close to the Sun might seem sufficient to wipe the comet out of existence, but there's a mitigating circumstance. The comet starts to melt and evaporate and, in no time, is surrounded by a cloud of steam. What's more, the comet is not just ice; it is dirty ice, containing large quantities of rocky grit. The cloud is therefore composed of steam and dust. This reflects much of the Sun's light and shades the comet, which thus manages to move past and begin its retreat still largely intact.

Still, the comet is badly affected by the approach. Some really sizable comets that approach the Sun fairly closely send out clouds of steam and dust that are pushed outward, away from the Sun, in a long tail. In 1843, there was a comet that formed a tail that was 190 million miles long, causing it to stretch from the vicinity of the Sun to out beyond the orbit of Mars. It obviously lost a lot of its mass in just that one fly-by.

Some time in the past, a sun-grazer must have been so damaged by the Sun's heat as to break into pieces. Indeed, the eight sun-grazers that have been detected in past years by telescopes were probably all pieces of the same original comet. They all followed just about the same orbit.

Undoubtedly, there are smaller pieces that have been formed, pieces too small to be detected until they get close enough to the Sun to form clouds of vapor, and then they are invisible in the Sun's glare.

But now there is a satellite called *Solar Maximum Mission* or, for short, *Solar Max*, which is designed to study the area near the Sun. It can do this because it possesses a *coronagraph* that blocks out the disk of the Sun so that the portion of the sky near it can be seen.

In October 1987 *Solar Max* photographed two streaks near the Sun that looked like small comets with tails extending outward away from the Sun. They passed into the region that was blocked off by the coronagraph, so that it could be assumed they were moving around the far side of the Sun and eventually they would be seen emerging on the other side of the blacked-out region.

But, it was announced in July 1988 that they never did. Assuming that their emergence had not been missed for one reason or other, the conclusion we must come to is that they were totally vaporized. It's also possible that they actually passed so deeply into the Sun's atmosphere that their orbits degenerated and they spiraled into the Sun itself.

Presumably, this happens now and then, and even more frequently for still smaller objects such as those we would call meteorites if they struck the Earth. Nevertheless, this dire fate is reserved only for small objects with lopsided orbits. Provided the solar system is left to itself, true planets with nearly circular orbits, like the Earth, are quite safe. They will not hit the Sun.

The Invisible Cloud

There is a part of the solar system that no one has ever seen but that nearly every astronomer is sure is there. In July 1987, three Soviet astronomers offered arguments that the invisible portion is much larger and more important than had been supposed.

Their theory started with comets. There are always comets streaking through the planetary system. Where are they all coming from?

Back in 1950, a Dutch astronomer, Jan Hendrik Oort, suggested that way out beyond the farthest known planet there is a vast cloud of small icy objects. Each, he theorized, slowly circles about the Sun in an orbit that takes millions of years to complete, and there may be billions of these objects altogether.

Every once in a while, something—a collision with another piece of icy debris or the gravitational pull of a nearby star—causes an icy object to slow and fall toward the Sun. It makes its way among the planets, and as it approaches the Sun, its ice vaporizes and the rocky dust within the ice lifts off the surface and forms a fog about the object. This fog is swept back by the solar wind into a huge tail, and the object has become what we call a *comet.* It sweeps around the Sun and then out toward the faroff cloud again.

Every once in a while, though, one of those comets is caught by the gravitational pull of a planet and then, like Halley's comet, it remains among the planets forever. It becomes a *short-period comet*, returning to the neighborhood of the Sun every few years or decades.

How large is this "Oort cloud" of distant comets? To estimate that, we must have some idea how large a typical comet is. Rocket probes were sent to pass near a comet for the first time during Halley's recent close approach, and they took certain measurements. It turned out that Halley's comet was considerably larger than had been suspected. It is an irregular object, but its average diameter is about 7.5 miles, and it contains about 140 cubic miles of ice. This amounts to about 30 billion tons of ice—quite a snowball.

The Soviet astronomers presented reasons for supposing that Halley's comet is a typical comet and that the Oort cloud is made up of objects averaging 30 billion tons in weight.

Recent estimates indicate that the thickest part of the Oort cloud lies at a distance of 2 trillion to 4 trillion miles from the Sun. That's roughly one thousand to two thousand times as far away as the most distant known planet, which is why the objects can't be seen. They are too far off. The most recent estimates of how many cometary objects may exist in this cloud is about 2 trillion (2,000,000,000,000).

If there are that many objects, each one with a mass equal to that of Halley's comet, the total mass of the Oort cloud is about one hundred times that of the Earth. This means that the total mass is roughly equal to that of Saturn, the second largest planet. This is about one thousand times as great as had been previously estimated, making the cloud a considerably more important portion of the solar system than had been thought.

Here is something else: Every object in the solar system rotates on its axis, and every object, except the Sun itself, revolves about the Sun. All this turning of an object about itself and other objects is measured as *angular momentum*, an important property of all objects from stars to electrons. There are two factors that determine how large angular motion might be: the mass of the object and the distance of the object from the center about which it turns.

The Sun is one thousand times as massive as all the planets and other objects that circle it put together, so you might think the Sun has almost all the angular momentum in the solar system. However, the Sun just turns about itself. Its various parts are not very far from its center, only 430,000 miles or so at most.

The planets, though much lighter than the Sun, move in grand sweeps that place them hundreds of millions of miles from the Sun. The distance more than makes up for the lightness of the planets. The result is that the Sun has only 2 percent of the angular momentum of the solar system. The planets have the other 98 percent.

In fact, although Jupiter is the largest planet with only a thousandth the mass of the Sun, it has about thirty times as much angular momentum.

But what about those comets, which are individually tiny but turn at a distance of trillions of miles from the Sun? The Soviet astronomers calculate that the comets have ten times as much angular momentum as all the rest of the solar system combined. This means that 90 percent of angular momentum is in the comets, 9.8 percent in the planets, and 0.2 percent in the Sun. If this is so, we may have to rethink our notions of the solar system's beginnings.

For the last forty years, scientists have worked out how angular momentum was transferred from the Sun to the small planets when the solar system was formed. It wasn't easy to do, and if they have to figure out how all that angular momentum was transferred to the distant Oort cloud, it will make things much harder.

What's in a Name?

Sometimes scientists, being human, become entangled in rather trivial disputes. Right now, for instance, some astronomers are getting a little heated about whether to call Pluto a planet or an asteroid.

Pluto was discovered in 1930 and was found to be circling the Sun at an average distance greater than that of any other planet. No one doubted that it was a planet, and it's been called that for over fifty years. The catch lies in its size.

When it was first discovered, it was thought to be somewhat larger than Earth, but it was so distant that it could only be seen as a dot of light and its actual size couldn't be measured. Little by little, though, information about it was gathered, and the more astronomers learned, the smaller Pluto proved to be. In recent years, Pluto was discovered to have a satellite, Charon, and when the Pluto-Charon system happened to move in front of a star, their sizes could be measured quite well.

We now believe that Pluto's diameter is about 1,420 miles, which is only three-fourths that of our Moon. Because Pluto is made of light icy material, it has a mass only one-sixth that of our rocky Moon. Some disgruntled astronomers, therefore, maintain that Pluto is too small to be considered a planet and should be demoted to the rank of asteroid.

Actually, there are three kinds of bodies in the solar system.

First, there is the Sun, which is so enormous (333,000

times the mass of Earth) that it supports hydrogen fusion at its center and glows with light and heat.

Second, there are planets, which are dark bodies that circle the Sun.

Third, there are satellites, which are dark bodies that circle the planets.

There is no chance of confusing these three types of objects. An object is either a sun, a planet, or a satellite, and we can tell at a glance which is which.

Planets, however, have a wide range of sizes. This was brought home to astronomers in the first decade of the 1800s when four planets, considerably smaller than any of the others, were discovered. They all circled the Sun between the orbits of Mars and Jupiter. Since then, thousands of other small planets have been discovered in the same region.

These small planets came to be called *asteroids*, meaning "starlike," because they were so small that they looked like dots of light in the telescope just as stars did, instead of expanding into circles of light as the larger planets did.

The asteroids *are* planets, however. They circle the Sun just as the other planets do, and the matter of size is secondary. In fact, it is possible to speak of "major planets," referring to the large ones, and "minor planets," referring to the small ones. There's no great astronomical need to make this division; it arises only out of the human habit of making neat classifications. If we do make the division, however, the question of where we should draw the line between major and minor arises.

Before the discovery of the asteroids, the smallest planet known was Mercury, which has a diameter of 3,013 miles, about two-fifths that of Earth and only about one-thirtieth that of the largest planet, Jupiter. Mercury is a small world, but it has always been called a planet and no one has ever suggested it was anything else.

As for the asteroids, or minor planets, the first one

discovered (January 1, 1801) is also the largest. It was named Ceres and has a diameter of 640 miles. Ceres is only a little over one-fifth as wide as Mercury and has no more than one-two-hundredth of the mass of Mercury.

There is quite a gap between Mercury and Ceres, you see. Until just a few years ago, it seemed quite fair to say that a major planet was the size of Mercury or more and that a minor planet (or asteroid) was the size of Ceres or less. In 180 years, no planetary object had been found that confused matters by existing in the gap between Mercury and Ceres.

And then the size of Pluto was finally determined. If its diameter is roughly 1,420 miles, then this is about two and a half times the diameter of Ceres. However, Mercury has a diameter just over two times that of Pluto. In terms of mass, Pluto is perhaps sixteen times as massive as Ceres, but Mercury is about sixteen times as massive as Pluto.

In short, Pluto falls just about midway between Mercury and Ceres. Well, then, on which side of the boundary line should it be considered? Should we consider it a major planet or a minor planet (asteroid)? It could go either way.

It really wouldn't matter which, but to prevent the astronomers from quarreling, I have a suggestion: Why not call any planet that lies between Mercury and Ceres in size a *mesoplanet,* since *meso* in Greek means "intermediate"; Pluto would be the only mesoplanet known at the moment—and wouldn't that make sense?

Pluto and Charon: The Dumbbell Worlds

The least-known world in the solar system is Pluto, but we are now learning some interesting things about it through a most unusual stroke of luck.

In 1978, an astronomer, James W. Christy, discovered that Pluto had a satellite, which he named Charon, after the boatman in the Greek myths who ferried dead souls across the River Styx to Pluto's underground domain. Every 124 years, Charon enters a five-year period during which, as seen from Earth, it passes directly in front of Pluto, then behind it, making a complete circuit over 6.4 days. It goes through this period of regular eclipses when Pluto is farthest from the Sun and again when it is nearest.

It happens that Charon was discovered just before it was to begin its five-year period of eclipses, and astronomers are still watching the effects avidly. What's more, Pluto is at the near point in its orbit right now, meaning that it is in the best position to be studied from Earth. If Charon had been discovered only five years later, astronomers would have lost their chance and would have had to wait two and a half centuries for the next near-point eclipses (though long before then we would have sent out probes to Pluto, to be sure).

The first thing astronomers have been able to learn during the eclipses is the size of the two bodies. By measuring the time Charon takes to cross the width of Pluto at the speed the satellite is known to be going, they can reckon the sizes of Pluto and Charon. Astronomers were able to determine that Pluto is only about 1,420 miles across. This makes it the smallest of all the planets. In fact, it is smaller than the seven largest satellites in the solar system. It has no more than one-tenth the mass of our own Moon, for instance. It wouldn't be fair to call Pluto an asteroid, however. Pluto is an in-between body: very small for a planet, but very large for an asteroid.

Charon is smaller still, of course—only eight hundred miles across—and has a diameter that is just over half that of Pluto, making the Pluto-Charon combination the nearest thing to a double-planet we know of. Until the discovery of Charon, the Earth-Moon system was the nearest to a double-planet, but the Moon has only one-fourth the diameter of the Earth.

When two worlds are close together, tidal effects slow their rotations. Thus, the Earth's tidal effect has slowed the Moon's rotation to the point where it shows only one hemisphere to the Earth as it circles us. Earth's rotation is also slowing because of the Moon's tidal effect, but Earth is so large that the slowing effect has been only partial so far.

Pluto and Charon, however, are only 12,250 miles apart, only one-twentieth the distance between Earth and the Moon, and that greatly increases the Pluto-Charon tidal effect. Pluto and Charon are, in addition, so small that they are more easily and quickly slowed. The result is that both worlds' rotations have slowed to the point where each shows only one hemisphere to the other. They face each other permanently and turn about each other as if they were the two halves of a dumbbell. They are the only two worlds in the solar system that turn about each other in this fashion.

During the eclipses astronomers have a chance to learn more about what Pluto and Charon are made of by studying the infrared light that they reflect. When Charon is behind Pluto, we see only the reflected light of Pluto. When Charon emerges from behind Pluto, we see the reflected light of both, and if we subtract the reflection of Pluto, we get the reflected light of Charon only.

From this reflected light, astronomers at the University of Arizona began, in March 1987, to work out the chemical nature of the surface of the worlds and of their atmospheres.

They have discovered that the surface of Pluto seems to be rich in methane, a substance that, on Earth, is a major part of the natural gas we use as fuel. Methane freezes at a very low temperature, so that even at Pluto's temperature, which may be −400 F (−204 C), some of it will evaporate and become a gas. It would seem, then, that Pluto has an atmosphere of methane gas that is about one-nine-hundredth as dense as Earth's atmosphere (or nearly one-tenth as dense as the thinner atmosphere of Mars).

Naturally, the temperature is lower at Pluto's poles, so there is more frozen methane there. Pluto may have polar icecaps of methane that become larger as it moves farther from the Sun.

Astronomers were surprised to find that Charon's reflected light was quite different from Pluto's. Because Charon is smaller than Pluto, it has a smaller gravitational pull. It can't hold on to the molecules of gaseous methane as well, and the methane has escaped during the billions of years that the solar system has existed.

What's left is frozen water, which doesn't vaporize at Charon's frigid temperatures and therefore isn't lost. Consequently, whereas Pluto has a methane surface and a very thin methane atmosphere, Charon is icy and has no significant atmosphere.

Before Charon's discovery in 1978, astronomers couldn't have imagined that they would find out so much detailed information about distant Pluto so soon.

The Case of the Missing "Planet"

Science has its disappointments. Every once in a while, a discovery that seems quite satisfying and points the way toward more is made—and then fades out. Too bad!

For instance, any object that is large enough, say with at least one-tenth the mass of the Sun, would, when forming, grow so hot at the center, and exert so much gravitational pressure there, that its central atoms would break down and begin fusing together, producing enormous quantities of radiation. In other words, an object that is large enough undergoes "nuclear ignition" and becomes the kind of cosmic hydrogen bomb we call a star. The greater the mass, the larger, hotter, and brighter the star.

Jupiter, the largest planet we know, is only one-thousandth the mass of the Sun. It isn't heavy enough to undergo nuclear ignition at its center, so it doesn't shine. We see it only by the reflected light of the Sun. If it were alone in space with no star nearby, it would be totally dark. It would be a *black dwarf:* black because it didn't shine, and dwarf because of its small size.

We've never located planets circling other stars. For one thing, the light they reflect would be very dim at the great distance of other stars. For another, that dim light would be drowned out by the brightness of the nearby star they circle.

Suppose, though, some star had a planet that was as much as fifty times the mass of Jupiter. This would still not be quite enough to induce nuclear ignition, but its interior might be hot enough to cause the planet's surface to radiate quantities of infrared light and even a little bit of dim visible light. This might not be much, but it would leave the object more detectable than if it shone only by reflected light. Such an object, in between the sizes of a large planet and a small star, might be called a "brown dwarf": not quite black.

In 1985, an object was detected very near the small star Van Biesbroek 8 (VB 8). VB 8 was dim enough, but the new object was even dimmer and the light it emitted was mostly in the infrared, which is less energetic than visible light. Indeed, the light was exactly what one would expect of a brown dwarf, and the astronomers who first detected it at Kitt Peak Observatory in Arizona were sure that that was what they had observed. They named it Van Biesbroek 8B (VB 8B).

There was some discussion of whether VB 8B ought to be called a very large planet, about fifty times the mass of Jupiter; or a very small star, about one-twentieth the mass of our Sun. The tendency was to consider it a very large planet. If that was what it was called, then it would be the very first planet detected that was circling a star other than our Sun.

The excitement was this: Now that a brown dwarf (a completely new type of celestial object) had been discovered, the same techniques would perhaps turn up many others. Studying such objects might give us new insights into what goes on at the centers of massive bodies, and we might understand all stars, including our own Sun, better.

In fact, it was even possible that there might be so many brown dwarfs in the universe that they would solve another puzzle. The stars we see seem to constitute only 10 percent of the mass that the universe apparently has. Perhaps the other 90 percent is composed of brown dwarfs.

Unfortunately, after the discovery of VB 8B, no further discoveries of that sort were made. Perhaps this was to be expected. They were just borderline objects, very difficult to see, and it might be that our astronomical instruments are not quite up to the task. A little more advance and brown dwarfs might be detected in every direction. Perhaps.

Then something much worse happened. In summer 1986, the discoverers took another look at VB 8B and found they couldn't locate it. A second group, working with an infrared telescope on Mauna Kea in Hawaii, also could not find it.

What happened? Of course, VB 8B might have moved. If it were a planet circling the dim star VB 8, it would be orbiting about it, just as Jupiter moves about the Sun. In this case, it might be that in the time since it was sighted, brown dwarf VB 8B had managed to move behind the star VB 8, or at least so close to it as to be lost in the glare.

In order to do that in the time since it was discovered, however, it would have to be quite massive. (The more massive an object, the stronger the gravitational pull between it and another massive object, and the faster the motion of one about the other.) In fact, it would have to be so massive that it would have attained nuclear ignition and it would be shining like a star.

That can't be right—so what other answer might there be? Well, for some reason, it could be that there was something wrong with the original observation and VB 8B just doesn't exist. And that's a disappointment more massive, almost, than the object itself.

[NOTE: Since this essay was written, other brown dwarfs have been reported; see page 348.—I. A.]

The Falling Moon of Mars

There must be many children who look at the Moon and wonder why it doesn't fall down.

Well, it won't. Quite the contrary. It's moving away from us. There are other moons that are falling, however. Toward the end of 1988, three British astronomers at an observatory in the Canary Islands made measurements of the motions of Phobos, one of Mars's moons, that put the matter beyond dispute.

Suppose we consider our own Moon first. The Moon moves in an orbit about the Earth, and if it were a perfect sphere and the Earth were a perfect sphere and there was no interference from outside, the Moon would stay in its orbit without change for an indefinite period.

However, the Moon pulls at the near side of the Earth more strongly than at the more distant far side, and this difference in pull creates the tides, and is spoken of as a "tidal effect." The Moon's tidal effect causes a bulge to appear on opposite sides of the Earth's surface.

The Moon pulls at that bulge, and the bulge pulls at the Moon. However, the Earth rotates on its axis in one day, while the Moon goes around the Earth in 27.33 days. This means that the bulge tends to be dragged along by Earth's rotation so that it is always just slightly ahead of the Moon.

This means that the Moon pulls backward on the bulge, tending to slow Earth's rotation, while the bulge pulls forward on the Moon, tending to speed it up.

The effect is very tiny but it can be measured. Because

of the tidal effect, Earth's day becomes a second longer every 62,500 years. This is not going to affect us noticeably in our lifetime, or even in the entire duration of civilization so far, but it does mount up.

Four hundred million years ago, the day was only 22 hours 13 minutes long, so that there were 395 days to the year. (The length of the year isn't changed by the tidal effect.) Fossilized coral remains have proved that this was the case. Since the calcium deposits in coral grow daily, and grow faster by day than by night, and faster in summer than in winter, they produce something like tree rings and the 400-million-year-old fossils demonstrate the shorter day unmistakably.

In the same way, the Moon, which is continually forced to move a bit faster by the bulge's pull, has an orbit that bellies outward because of this faster motion. After each revolution of the Moon, it is about .1 inch farther from the Earth. This is not large enough to notice from revolution to revolution, but it mounts up, too.

For instance, the Moon has a disk, as seen from Earth, that is just about as large as the Sun's disk, as seen from Earth. This means that every once in a while, the Moon moves in front of the Sun (as seen from Earth) and we see a beautiful total eclipse. However, as the Moon moves away from the Earth, its apparent disk decreases in size, while the Sun's disk does not change.

In about 750 million years, the Moon will appear sufficiently small so that there will never be a total eclipse of the Sun; the Moon's disk will never totally cover the Sun. Still, I suppose you have to take a really long view to worry about something like that.

But what about Phobos, Mars's nearer satellite? It is a small potato-shaped object about 17 miles in its longest diameter. It circles Mars only about 5,840 miles above its surface. It, too, produces a bulge on Mars's surface through a tidal effect. Because Phobos is so much smaller than our Moon, it produces a smaller bulge and has only a small effect on Mars. But the tiny bulge on Mars has a large effect on the tiny satellite.

Mars turns on its axis in 24.5 hours. Phobos, however, is so close to Mars (much closer than our Moon is to us) that it revolves about Mars in only 7.65 hours. Phobos races ahead of Mars's surface, rising in the west and setting in the east. Because it races ahead, it tends to be slightly ahead of the bulge it produces, so that its gravitational pull speeds Mars's rotation very slightly, while the bulge pulls back on Phobos and slows it.

As Phobos's rotation slows, it drops closer to Mars. Every year, it moves 1.5 inches closer to Mars and its time of rotation decreases by a few hundredths of a second. The measurements in the Canary Islands in late 1988 show that in the past ten years, Phobos has moved 14 inches closer to Mars.

The closer it moves, the larger the bulge becomes and the more rapidly Phobos loses altitude. Eventually, as it comes closer to Mars, Mars's intensifying gravitational field will tear Phobos into fragments that will rain down on the planet. Phobos has been circling Mars for billions of years, perhaps, and we now have the exciting chance to see it in the very last stages of its life.

Of course, even the last stages, short to an astronomer, would be long to anyone else. It will still take about 38 million years before Phobos breaks up and falls, so don't hold your breath.

Life on Mars Revisited

Possibly, just possibly, we have discovered organic matter in the Martian surface and, therefore, the faintest possible hope that there may be life or that there may once have been life.

Back in 1976, the United States placed two Viking probes on the Martian surface. The probes scooped up Martian soil and ran tests that, it was hoped, would indicate whether microscopic life was present. Several of the tests were ambiguous. Scientists could not tell, from those tests alone, whether life was present or whether the results were due to some unusual nonlife chemistry.

One of the tests, however, seemed to indicate there was no organic matter in the soil, no carbon-containing material. Because life, as we know it, is entirely based on carbon-containing material, there is no life without it. It was decided, therefore, that Mars was almost certainly lifeless. But now scientists have had another look. No new probes have landed on Mars to investigate, but some of Mars may have come to us.

It came about this way. In the last dozen years or so, scientists have been picking up meteorites in Antarctica. In most places in the world it is very difficult to tell whether a meteorite has fallen unless it is actually seen to fall. Once it has landed, it looks very much like an ordinary rock unless it is carefully tested chemically, and it would be extremely difficult for anyone to take a close chemical look at all the rocks that litter the Earth.

On the vast Antarctic ice cap, however, there is only ice. If any piece of rock is found upon it, it can only have been deposited there as a meteorite. As a result, scientists have now collected a considerable number of Antarctic rocks they know to be meteorites. What's more, meteorites that land almost anywhere else on Earth are degraded by liquid water and invaded by microscopic life. On lifeless Antarctica, containing water only in the frozen state, the meteorites are not touched but remain exactly as they were when they landed.

A few of the meteorites have the same composition as do the Moon rocks brought back by the astronauts. The feeling is that the bombardment of the Moon that produced its craters may have splashed some bits of the Moon's surface into space and sent them flying to Earth. A few Antarctic meteorites include bits of gas that have the precise composition of the Martian atmosphere, and these, many astronomers are convinced, may have come from Mars.

One of these meteorites was carefully analyzed earlier this year by a team of British astronomers under Ian P. Wright. In it, small amounts of two different kinds of carbon-containing compounds were found. One consisted of bits of calcium carbonate, ordinary limestone. The other, however, consists of organic compounds, whose exact nature has not yet been identified, but which are probably related to the kind of material found in living tissue.

It follows that if the meteorite did indeed come from Mars and if it is representative of the Martian surface, then there must be organic compounds in that surface, regardless of the tests conducted by the Viking probes. After all, the Viking probes landed on two very tiny isolated points of a vast planetary surface, and they may just have landed in places that happened not to contain organic matter. And if there is organic material in the Martian surface, then some form of life, possibly very primitive, may exist there now, or may have existed in the past.

But even if the meteorite came from Mars, is that where the organic matter came from? After all, the meteorite arrived on Earth only because Mars received the impact of a colliding body that drove pieces of Mars out into space. The colliding body may have been a comet, and comets are known to be made up, in part, of carbon-containing compounds. In that case, the meteorite may be from Mars, but the organic material may be from the comet.

Carbon exists in two varieties, carbon 12 and carbon 13. The relative proportions of the two are slightly different on Earth and in cometary materials. Some astronomers point out that the ratio in the meteorite is not what one finds in comets but is characteristic of Earthly origin. Were the meteorites contaminated, somehow, while they were being handled by the scientists who collected them, preserved them, and eventually analyzed them?

Wright and his team, however, argue that the meteorite was handled too carefully to be contaminated. If the carbon 12/carbon 13 ratio is wrong for cometary origin, and could not have arisen from the Earth, then this would seem to be further evidence that the carbon-containing material must have originated on Mars.

And why did the Viking probes not detect such material? Wright claims it wasn't chance. The probes scooped up material from the very top surface layer of Mars. A comet striking Mars, however, would plow up the Martian ground and send material streaking to Earth from farther down, where the carbon-containing compounds may be concentrated. It is a pretty fascinating problem, and won't easily be settled.

A Little Brighter

There is something out beyond Saturn that has been puzzling astronomers for a dozen years. It is some sort of heavenly body, but exactly what kind was unclear. Now, its true identity may finally be emerging.

The story started on November 1, 1977, when the American astronomer Charles Kowal discovered what seemed to be an asteroid that was moving slowly, very slowly. The more slowly an asteroid moves, the farther it is from the Sun, and this one was farther than any asteroid ever seen, for it circled the Sun beyond the orbit of Saturn.

The only small objects that had ever been seen as far as Saturn or beyond were the satellites that circled the distant planets: Saturn, Uranus, Neptune, and Pluto. What Kowal had discovered was a small object that moves in an independent orbit around the Sun, moving around it sometimes at about the distance of Saturn's orbit and then moving out as far away as the orbit of Uranus. Its orbit is tilted in such a way, however, that it stays far below or above these two planets as it moves through its own orbit. There is no danger of a collision.

Kowal looked for it in old photographs of appropriate portions of the sky and worked out its orbit. It circled the Sun every fifty-one years. Its orbit carried it to within 790 million miles of the Sun at one end and out as far as 1.74 million miles at the other end. Because it seems to gallop endlessly closer to and farther from the orbits of Saturn to

Uranus, Kowal named it Chiron after the most famous centaur (half-man, half-horse creature) in the Greek myths.

The question of what it might be arose. It might be an asteroid. It is fairly large, for it is 112 miles across, but there are asteroids of that size known. The only trouble with this notion is its distance from the Sun. All the asteroids we know have all or part of their orbits in the space between Jupiter and Mars (the *asteroid belt*). A few tiny asteroids with orbits inside that of Mars are known, but Chiron would be the only asteroid we know with an orbit that lies entirely beyond that of Jupiter.

Of course, the more distant an asteroid, the harder it is to see. Perhaps the outer solar system beyond Jupiter is littered with asteroids, which are so far away we can't see them from Earth. Perhaps we can barely make out Chiron because it is unusually large for an asteroid. Perhaps when the day comes that we have telescopes in orbit far out in space, we will discover many more Chiron-like objects.

On the other hand, Chiron may be a comet. Comets are known to exist far out in the solar system. Chiron is large for a comet, to be sure, some two thousand times as massive as Halley's comet, for instance, but perhaps some comets are that large.

Chiron showed no signs of being a comet, however. There is this difference between an asteroid and a comet: An asteroid is made up, mostly or entirely, of rocky or metallic materials that do not vaporize even when red-hot. A comet is made up mostly of icy materials that vaporize when heated, forming a dusty cloud about itself. This is why comets, as they near the Sun, grow fuzzy and develop a long tail.

Chiron showed no signs of fuzziness, but this may be because it is so far from the Sun that it receives insufficient heat to vaporize its ice. Chiron was, however, near its maximum distance from the Sun when it was discovered in 1977, and it has been approaching ever since. It will reach its minimum distance in 1996.

This means that since it was discovered, it has been edging closer and closer to the Sun and getting warmer and warmer.

Naturally, as it gets closer to the Sun, it receives and reflects more light so that it gets brighter. Astronomers have a pretty good idea of how an asteroid would brighten as it approaches the Sun, and even as early as November 1987 it seemed that Chiron was becoming just a little brighter than it ought to be.

Now, Karen J. Meech of the University of Hawaii and Michael J. S. Belton at Kitt Peak Observatory in Tucson report further brightening that can only be the result of sunlight reflected from an atmosphere of vapors developing about Chiron. This would seem to mean that Chiron is not an asteroid but a giant comet after all.

Perhaps it is not unusually large for a comet. Perhaps very many of the comets that are thought to exist far beyond the orbit of Pluto are this large. The ones we see close-up, after all, are those that come into our own vicinity, very near the Sun, over and over again. Each time they approach the Sun, much of their substance vaporizes so that they are now much smaller than they once were.

If an object as large as Chiron had its orbit altered by planetary pull and were made to drop into our section of the solar system, it would lose so much vapor that it would develop a giant cloud about itself that would be larger than the Sun and a tail that would be hundreds of millions of miles long and stretch halfway across the sky. Several such giant comets were observed in the 1800s, but in our own century, alas, we have seen only puny examples. We can only stare at Chiron and think of the sights we are missing.

Space Pollution

In recent years, I have given talks in which I stressed that space was so huge in volume that we need not fear that we would pollute it with our activities. How wrong I was! It took ten thousand years of civilization before we began to pollute the whole of Earth's ocean, soil, and atmosphere in a significant way, but it has taken us only thirty years to pollute space in Earth's neighborhood.

We have put objects into space by the thousands in those thirty years. If these objects were to remain motionless with respect to Earth's surface, they would do no significant harm in remaining where we put them; there would be plenty of room, for the volume of space is indeed enormous. However, if they were motionless, they would all fall to the ground. They remain in space only because they are all racing about the Earth at speeds of up to five miles per second. At these speeds every object in space is a bullet and, in most cases, far more dangerous than the bullets we fire out of guns.

There are about three hundred operating satellites orbiting Earth right now, but there are many more satellites that, though they have ceased operation, are still whirling about up there.

Nor are the satellites themselves all there is. These satellites were hurled into space by rockets, and there are pieces of rocketry that are still in space as a result.

Some satellites have exploded or have collided, one with another, and each time this happens, they fragment into small pieces, all of which continue to orbit the Earth.

There are, as a result, six thousand man-made pieces of

debris large enough to be tracked by radar, and they are being tracked. There are, however, many more bits of matter that are too small to be tracked. According to some estimates, there are sixty thousand pieces of debris about an inch in size. There may also be uncounted millions of flecks of paint.

We may smile at the thought of engineers becoming upset over a fleck of paint, but even such an inconsiderable object becomes something to worry about when it is traveling at a rate of several miles per second. In June 1983, a fleck of paint that was only one one-hundred-twenty-fifth of an inch across—too small to see—struck a window of the space shuttle *Challenger.* The collision managed to gouge out a bit of glass, leaving a tiny crater, one-tenth of an inch across, in that window. This may not seem like much, but it weakened the window sufficiently to make it necessary to replace it, at a cost of $50,000, before the shuttle flew again. That was an expensive fleck of paint, then, and if something a little more massive had struck, there might have been a disaster on the *Challenger* two and one-half years earlier than the explosion that killed seven crew members.

And the situation is growing worse. The United States, the Soviet Union, and other nations are continuing to launch objects into space. Explosions and collisions continue. The amount of debris continues to mount, so that some people estimate that the number of pieces in space will quadruple every ten years.

This means that it is quite possible that, by the year 2000, we can expect that any working satellite in any given year has one chance in two hundred of being hit by a piece of debris about one inch across. If there are four hundred working satellites in space at that time, then we can expect, on average, that each year two working satellites will be struck. The damage is quite likely to be serious; if the debris happens to strike a particularly vital part, the satellite may be put out of action altogether.

Satellites will have to be made more rugged if they are to survive; this means that they must be made more massive and therefore more expensive to launch. And what about spacesuits and spaceships? They are not 100 percent safe either. It is possible that in one hundred years, it simply won't be safe to try a "space walk" in the neighborhood of Earth. Eventually, it may well come to pass that space will be so full of debris that space flight through Earth's permanent ring of space garbage will become an increasingly hazardous undertaking.

What are we going to do about this? We might try to cut down on the number of satellites we launch or take measures to prevent as many explosions and collisions as possible—and certainly stop any project that involves the deliberate destruction of satellites.

But that only slows the increase of danger. It doesn't remove it. Ideally we ought to think up some way of periodically cleaning up space, of running a vacuum cleaner over it, so to speak. Unfortunately, there doesn't seem any readily available way of working up an efficient and affordable cleaning system.

Where Do We Go from Here?

Having gotten back on track with the space shuttle of 1988, where do we go from here? It's important to have our future in space planned, for the road is an expensive one and we cannot afford to flounder.

One obvious dream goal is that of a manned flight to Mars and its satellites. If we accomplish this, we will explore a world that is not too far away and that in some ways is like the Earth. It is smaller and colder, but it has a thin atmosphere, a twenty-four-hour day, and ice caps. And it has mysteries, too: dried-up riverbeds that once may have flowed with water, volcanoes that once may have spewed lava, a vast canyon that may betoken a once-active crust.

Yet the task of sending human beings to Mars and bringing them back alive is so enormous and so barely within the realm of possibility that neither the United States nor the Soviet Union can undertake it without backbreaking effort and unimaginably suspenseful fears for the safety of the astronauts. It becomes marginally less dangerous if the United States and the Soviet Union pool their resources and expertise, making the Mars project a global effort rather than a national one. This might encourage globalism in other directions, too, and because the problems we now face on Earth are global in nature and require global solutions, this might be an even happier result of this difficult project than the exploration of Mars would be.

Still, a trip to Mars from Earth as base is bound to be a showpiece not easily repeated. It would be like the trips to the Moon fifteen years ago, which, however spectacular, seemed to lead to nothing broader and deeper.

It is absolutely necessary to our ventures into space that we build a base away from Earth, a base with a lesser gravity and without an interfering atmosphere.

The logical beginning is with a space station, one larger and more versatile than the Soviets have set up in space, one that would be continuously inhabited by crews working in shifts. The parts for constructing new space vessels would be transported to the space station. The intact vessels could not be lifted off Earth without vast rocketry, but the parts could be taken up much more cheaply and safely. Once built, the vessels would be taking off under weaker gravitational pull than they would from the more distant Earth and would have the initial kick of the space station's orbital velocity. They would need less fuel and would carry larger payloads.

With a space station as base, it would be far easier to reach the Moon and set up a permanent base there. The Moon could then serve as a huge mine. Suitable chunks of the Moon's surface could be fired into space by means of "mass drivers" that used electromagnetic forces for propulsion. This would be relatively easy on the Moon, where the surface gravity is only one-sixth that of Earth. In space the lunar ore could be smelted and from it all structural metals could be obtained, as well as concrete, glass, and soil.

It is with Moon materials that we will be able to build structures in space: power stations that make use of solar energy and relay it to Earth; automated factories that take advantage of the special properties of space and help lift the pall of industry and pollution from Earth itself; settlements that may each be large enough to house one thousand human beings in orbit about Earth under conditions that closely mimic the environment to which we are accustomed.

It may take us the better part of the twenty-first century to build up, and put into use, the space between Earth and Moon. But once this is done, we will have, at last, a firm base for operations in space beyond, one that is far superior to Earth itself.

The inhabitants of the settlements will be accustomed to space as Earth people can never be. They will be used to living inside an artificial world. They will be accustomed to changes in apparent gravitational pull as they move about their small worlds. They will take for granted the necessity to recycle tightly all the air, water, and food they use.

When a settler steps into a spaceship, he will be moving

into a world that is smaller than the one he is used to but whose properties will remain familiar. What would be impossibly foreign to an Earth person would be home sweet home to a settler.

The settlers, then, being much better suited psychologically to life on a spaceship, will be better equipped to face long voyages through space. It is they who will be the Phoenicians, the Vikings, the Polynesians of the future, making their way into the twenty-second century through a space ocean far vaster than the water ocean traversed by their predecessors.

It is from the settlements as base that repeated voyages to Mars and its satellites can be made. This will be only the start, too, for other trips can be made to the asteroids, to the satellites of Jupiter, and eventually to all the solar system. And beyond that are the goals of the twenty-third century—the nearer stars.

The End?

Eventually, everything must come to an end: you and I, all humanity, the Earth itself. But what will the end be? Scientists speculate on such things, and here's one possibility that is the latest scenario I could find.

Suppose we *don't* have a nuclear war. Suppose we solve all the problems that face us today. Suppose we learn how to improve the human body and mind, making ourselves stronger, healthier, and wiser. Can we then keep on going forever? Can we and our descendants continue to evolve, and to improve our beloved planet, and look forward to a perpetual Garden of Eden?

No, we can't. Our problem is the Sun. Unlike Earth, it is not a quiet and placid structure. Gravity has compressed the Earth as far as possible, and it will stay as it is indefinitely, if left to itself. The Sun, however, is huge, and its gravitation is capable of collapsing it to a pygmy. It does not collapse, but that is only because it is constantly generating heat at its center. This heat prevents it from collapsing from the pull of its own gravity.

This heat is generated because hundreds of millions of tons of the Sun's hydrogen atoms (which make up 75 percent of its mass) are being fused to more complex helium atoms every second. This fusion reaction generates heat, giving the Sun a large helium core that is growing steadily. The Sun contains so much hydrogen that even after nearly 5 billion years of fusion, there is still plenty left.

Still, everything must end eventually. In five or six billion years more, the Sun's hydrogen will be running low and its helium core will have become crucially large and hot. It will reach a point at which the helium atoms will start fusing to still more complex atoms. There will be a sudden flow of additional heat and the Sun will begin to expand. It will grow much larger and the outermost layers will become cooler. The surface of the Sun will decline from white heat to mere red heat and it will become a *red giant*.

Though the outer layers of the Sun will cool, its orb will grow so large that the total heat reaching Earth will steadily increase as the Sun expands. Long before the Sun reaches its maximum size, the Earth will be scorched and sterilized and there will be no life at all left on it.

How large will the Sun be when it reaches its maximum size?

The most recent calculations I have seen estimate that it will reach a diameter of a little more than 200 million miles. This means the Sun will fill the entire orbit of the Earth and a little beyond. The Earth will eventually be revolving about the Sun's center about 7 million miles below its surface.

Of course, this isn't quite as bad as it sounds. The outermost layers of a red giant are so thin that they're little more than a vacuum. Even though the temperatures of these outermost layers are still as high as 1,500 F (822 C), the small quantity of matter present won't produce enough total heat to melt the Earth. We might picture the Earth as a ball of indigestible rock and metal, circling amid the outermost wisps of gas of the Sun. Even though the Earth will be lifeless at this time, it may comfort us a bit to know that the world that was our home might still continue to exist. However, the Earth would only continue to do so if it were to remain in its present orbit and stay in that outermost layer of gas—but it won't.

The wisps of gas by which the Earth will be surrounded will still be thick enough to slow the Earth's motion very slightly and to cause it to spiral, very gradually, toward the Sun's center. The trouble with this is that as it sinks toward the center, it encounters thicker gas. The slowing of its motion becomes more pronounced and it sinks toward the center faster and faster. As it sinks toward the center, the temperature of its surroundings rises and there is more solar matter to transfer heat to the Earth. Within a few centuries, the Earth will grow hot enough to melt, vaporize—and vanish.

The Sun can continue to generate heat for more than 10 billion years, but once it becomes a giant its end is near. What fuel it has will decrease in a few million years to the point where too little heat is generated to prevent the Sun from collapsing. Gravity will then have its way at last, and the Sun will contract to a size smaller than that of the Earth.

Its surface layers will heat up again and it will become a *white dwarf.* About this small remnant, the outer planets will still circle, but Mercury, Venus, the Moon, and Earth will be gone forever.

But remember, this won't happen for billions of years, so there's plenty of time for humanity and its descendants to prepare, assuming they avoid other kinds of disaster. Surely, by then, humanity will have practical interstellar travel. We will find it easy to build huge cities in space that will carry us outward on long, long journeys to planets circling other, younger stars. We will look back on Earth and Sun in sorrow, but we may feel proud if we fragile human beings can survive the very Earth itself and the Sun that warms it.

Are We Alone?

One of the favorite games in science is to try to assess the chances of life's existing in the universe. Is Earth the only life-bearing planet? Or are there countless billions of life-bearing planets out there? Scientists have alternated between pessimism and optimism, but recently a new note of optimism has been sounded.

In the first third of the twentieth century, it was felt that planets arose out of a near-collision of two stars. This is so unlikely a phenomenon that there might well be only two sets of planetary systems in the entire galaxy, our own and that of the star that sideswiped us. In that case, pessimism had to be deep. Life would be so rare that we might really be alone in the universe.

Starting in 1944, however, new and far better analyses of the planet-forming phenomenon made it seem that every star might have planets of some type. In that case, optimism burgeoned and it was easy to suppose that life might be common. How common? That would depend on how stringent the conditions must be for life to form.

Considering our own solar system, there was a feeling at first that Venus might be warmer than Earth. However, thanks to its thick cloud layer, it was probably not too much warmer. Mars, it was believed, was colder than Earth but perhaps not too much colder. Therefore, the possibility of life would exist for any planet that was located about a sunlike star between the distance of Venus and Mars. This made for a broad "ecosphere" and increased the likelihood of life elsewhere. Optimism grew.

But then, with the coming of the age of rockets and probes, we had a chance to study Venus and Mars up close and, behold, Venus was far too hot to support life, and Mars was far too cold.

With this, there came a sudden increase in pessimism. Astronomers at NASA's Goddard Space Flight Center in Greenbelt, Maryland, made some calculations based on our new knowledge of our neighbor planets. They decided that if Earth were just 5 percent closer to the Sun (88 million miles instead of our 93 million), there would be a runaway greenhouse effect and Earth would become too hot to be habitable. On the other hand, if we were just 1 percent farther from the Sun (94 million instead of 93 million), the glaciers would take over. And if the Earth's orbit were a little more elliptical than it is so that it was too close in some spots and too far in others, it would constantly go from one extreme to the other.

This greatly narrowed the ecosphere. Clearly, it is just

the greatest stroke of luck that Earth has a nearly circular orbit that keeps it within that dreadfully narrow ecosphere at all times. The chance that this would happen in the case of other Sun-like stars is so small that again we are forced to think that there might be very few planets that are actually habitable. And even those that are might happen not to form life. Again, we might be alone in the galaxy.

Closer to the Sun, things continue to look bad. After all, Venus is almost the twin of Earth as far as size is concerned and it *is* much hotter than we thought it would be: hot enough to melt lead. This seems pretty conclusive.

If we look farther from the Sun, however, how far can we trust the case of Mars? To be sure, Mars is colder than Antarctica, but it is a small planet with only one-tenth the mass of Earth. This means that its gravitational pull can, at best, hold only a thin atmosphere and that it can also hold less internal heat.

There's no reason, however, to think that there is some rule that makes it necessary for a planet at Mars's distance from the Sun to be so small. It might just as well have been larger. Suppose, then, that Mars had formed at its present distance from the Sun but had happened to be as large as Earth. A planet as large as Earth, but farther from the Sun, would collect an even thicker atmosphere than we have and an ocean as well. The atmosphere might be mostly carbon dioxide and this, together with water vapor, would create a greenhouse effect that would keep Mars considerably milder than it is now. Its internal heat and volcanic action might also contribute.

Being warm enough to support life might not guarantee that life would form, and even if it did, the life might be completely different from that on Earth. For one thing, if Martian life replaced the atmospheric carbon dioxide with oxygen (as life on Earth did), then the greenhouse effect would be lost and Mars would cool down.

Nevertheless, astronomers at NASA's Ames Research

Center in California have reversed the pessimism of their fellow astronomers at Goddard, suggesting that the ecosphere might be expanded again—not to its full earlier limits of the distance between Venus and Mars, but at least through the cooler half, from Earth to Mars.

About 10 percent of the stars in our galaxy are reasonably Sun-like. With a broader ecosphere, it may be that half of them have a planet in that habitable band. This would mean 5 billion planets, at least, that may be habitable. But how many have actually formed life, and intelligent life at that—that's another matter.

V

FRONTIERS OF THE UNIVERSE

The Supernova Next Door

Astronomy isn't much of an experimental science. Astronomers can only watch the sky and take what it shows them. Sometimes it simply won't show them what they want to see.

For instance, between the years 1006 and 1604, five supernovas appeared in the sky. Five stars of our galaxy exploded in an unimaginably vast inferno, so that each one of them shone, for a few weeks, with the light of a billion stars like our Sun and then slowly faded in the course of months.

Ordinarily too dim to see with the naked eye, these stars that suddenly burst forth with the brilliance of Jupiter or Venus seemed like new stars. In 1572, one was studied by a first-rate astronomer for the first time. He was Tycho Brahe, and he wrote a book with a Latin title that, in its short form, is *De Nova Stella* ("Concerning the New Star"). After that, all exploding stars were called novas. Some explosions are minor, though. The really large explosions, like that of 1572, are now known as supernovas.

Tycho Brahe, however, didn't have a telescope. It hadn't been invented yet. The telescope wasn't used to view the sky until 1609, five years after the last of the five supernovas appeared.

Since 1609, we have had steadily larger telescopes, also spectroscopes, photography, radio telescopes, and computers: all the paraphernalia of high-tech astronomy. What we haven't had is a supernova. Since 1604, not one supernova has exploded in our galaxy.

That isn't to say that there haven't been any, just not in our galaxy. We see them in distant galaxies. The supernova of 1604 was about 35,000 light-years away, scientists estimate, but the closest one until a few years ago appeared in 1886 in the Andromeda Galaxy, which is 2.3 million light-years away, sixty-five times as far away as the 1604 supernova. What's more, astronomers didn't know the 1886 explosion was a supernova and didn't study it as carefully as they might have.

It was only in the 1930s that astronomers realized what supernovas were and started watching the sky for them. Since then, about four hundred supernovas have been detected, but all of them have been in distant galaxies that were many millions of light-years away, much farther even than the Andromeda supernova.

Does it matter? Yes, it does. Astronomers are trying to work out what goes on at the center of stars. If they could observe a supernova close up (not too close up, of course—say, just a few thousand light-years away), using today's advanced instruments, the actual details of the explosion might give us much better insight into events at the center than any we now have. It might enable us to understand our own Sun better.

But this accounts for the frustration of the astronomers. How about the general public? Why should they care about supernovas?

Well, for one thing, at the time of the big bang, when the universe began, the only substances formed were the elements hydrogen and helium. All other elements, without exception, have been formed at the center of stars and usually stay there forever. Supernovas, however, when they explode, spread the higher elements far and wide, and later stars, when they form, incorporate these elements.

The Earth is made up almost entirely of the higher elements. Ninety percent of the mass of the human body consists of elements other than hydrogen or helium. This means that almost every atom in us and in Earth was formed in a star that became a supernova.

Second, our solar system formed from a collapsing cloud of dust and gas. But what made it collapse when it had been sitting there quietly for billions of years? The best guess is that a supernova went off nearby, compressed the cloud, and started the collapse.

Third, supernovas produce vast quantities of cosmic rays and Earth is constantly being sprayed with cosmic rays that originated in various supernovas here and there in the sky. These cosmic rays produce mutations and accelerate the process of evolution. Without them we would all still be one-celled creatures—if that.

So supernovas are responsible for our existence in three different ways.

But in 1987 a flash of light reached us from a supernova that exploded in the Large Magellanic Cloud. It is not in our own galaxy, but it is at least in the nearest outside galaxy. It is only 155,000 light-years away, only four and one-half times as far away as the supernova of 1604 and only one-fourteenth as far as the Andromeda supernova.

It is the first chance astronomers have had to study a pretty close explosion of this sort and they are taking full advantage. Whatever they find, there are bound to be useful surprises that will extend our knowledge.

The Planet Hunt

For nearly half a century now, astronomers have been persuaded that planets must be common and must accompany most or all stars. This would be especially true of single stars, like our Sun, that don't have nearby companion stars. Now the astronomers finally have the most reliable evidence yet that this belief may be correct.

The present conception of the manner in which stars are formed involves a large dust cloud that slowly condenses, turning faster and faster as it does so. The central part becomes a star, but the thinner material surrounding it eventually produces planets. In fact, such condensation can't help but produce planets near stars; our own solar system is an example of this. The trouble is that it's the only example we know.

If there are planets circling other stars, we certainly can't "see" them in the ordinary sense of the word. A planet doesn't shine with light of its own but only by the light reflected from the nearby star it circles. It is therefore much dimmer than a star, and what little light it can give off is utterly lost in the glare of that nearby star.

But we needn't actually see a planet to know it is there.

A star without planets (or companion stars) tends to creep across our sky in a slow, perfectly straight path. If, however, a planet accompanies it, the planet and the star turn about a common center of gravity. The planet, being smaller, with less gravity, does most of the turning, but the star also turns in a small wobble. When this happens, the star follows a path across the sky that has a very slight wave to it. (Viewed from a distance, our own Sun would show a wobbly path, due chiefly to the pull of the large planet Jupiter.)

The wobble is most noticeable if the star is small and the planet is large. From the 1940s to the 1960s such wobbles were reported for some stars, notably a small one called Barnard's Star, which is only 5.9 light-years away.

The reports could not be confirmed, however. Other astronomers could not measure the reported wobble, and it was finally decided that the report arose from errors in the use of the telescope. Hope faded. In the last couple of decades, however, instruments have improved and, in mid-1988, two astronomers, David W. Latham of Harvard and Bruce Campbell of the University of Victoria in British Columbia, have each reported observing stars that wobbled.

Latham's discovery was more or less accidental. He was studying a Sun-like star known as HD 114762, about 90 light-years from us, simply to test his telescope, and he found a wobble. Not wishing to make another premature announcement, he kept an eye on the star for seven years, during which time (judging from the wobble) a planet circled it thirty times, with a period of revolution equal to eighty-four days.

Campbell, on the other hand, studied the manner in which stars approached (or receded from) us. With a planet, the star might wobble as it approached, then recede, then approach, in alternation. Out of eighteen stars studied by Campbell over a seven-year period, all within 100 light-years of us, nine showed a wobble, but if this was the result of the existence of planets, those planets were far enough away from the stars they were circling that they took longer than seven years to complete a revolution. Because an entire wobble was not observed, the results were somewhat less certain than Latham's.

For the wobble to be noticeable, the planets have to be large, probably considerably larger than Jupiter. This makes it uncertain whether they are truly planets or merely very dim companion stars. And, of course, even if they are planets, planets that are as large as Jupiter are bound to be made up mostly of hot hydrogen and would be completely unsuitable for life resembling our own.

Still, what these results seem to show is this: At least half the stars, and maybe more, have some sort of companions that are not too obviously stars. They may be Jupiter-type planets, but it seems likely that, if a Jupiter-type planet exists, there are other planets also circling the star that can't be detected simply because they are too small and light to force a detectable wobble on the star.

In other words, with these reports, astronomers are bound to be a little readier to believe that there are numerous Earth-like planets in our galaxy (and in other galaxies as well). This is important, because the more Earth-like planets there are, the greater the chance that at least some of them will have conditions suitable for life and that on them life will develop.

Thanks to this news, those astronomers who (like me) suspect that life may be a common phenomenon in the universe can now feel just a little surer of their ground. And if life is common, then, very occasionally, intelligent life-forms may develop and produce technological civilizations, so that we are not alone.

Far Beyond

In modern science, we now stand on the brink of projects that, like medieval cathedrals, will be begun by those who know they will not witness their completion.

So far, for instance, we have sent probes to the outer planets. *Voyager 2* has photographed Uranus and Neptune, the farthest known planets. The project has taken over a decade, but even middle-aged astronomers could expect to live a decade and view the end.

After leaving Neptune, *Voyager 2* will continue onward indefinitely, out beyond the familiar planets and through the void of interstellar space. It will serve no purpose out there, of course, and will merely be an unmarked wanderer.

Now, however, astronomers are speculating on the possibility of launching a probe that will be useful to us even when it is far beyond the outermost planet. It would leave the Earth at a comparatively low speed and will contain some twelve and one-half tons of frozen xenon. This will be heated until its atoms break up into electrically charged fragments (*ions*). The ions will be expelled forcibly, little by little, so that the probe will slowly accelerate for a period of ten years.

At the end of the ten-year acceleration, the xenon will all be gone, and by that time the probe will be moving at a speed of 225,000 miles per hour, or 62.5 miles per second. It will then be about 6 billion miles from Earth, well beyond the farthest point reached by that little, far-out planet, Pluto.

At that point the fuel tanks will be jettisoned and the probe itself, about five and one-half tons in mass, will continue to move outward at a speed that will very slowly decrease because of the weak pull of the distant Sun.

It will continue to drift outward for forty more years until it is nearly 100 billion miles away from the Sun. This is about a thousand times as far from the Sun as we are. The distance of the Earth from the Sun (93 million miles) is called an *astronomical unit* (AU). The distance of the probe after fifty years will be a thousand astronomical units (1,000 AUs), so it is called the TAU project.

The TAU probe will have a large telescope on board and its job will be to send us pictures of the stars taken at increasingly large distances, until the final pictures are taken 1,000 AUs away. After that, with the probe's energy supply gone, it will continue onward as indefinitely and uselessly as earlier probes have.

Of what use will such distant pictures of stars be?

When stars are viewed from different places, the nearer ones seem to shift position compared to the farther ones; this shift is called *parallax*. The greater the shift, the nearer the star. By measuring the size of the shift, we can calculate the distance of the star.

Unfortunately, even the nearest stars are so far away that the shift in position is exceedingly small. We can increase the shift by viewing the star from two places that are very distant from each other. On Earth, however, the farthest distance we can deal with is Earth's position in space at one time, and its position six months later when it is at the opposite end of its orbit. The extreme ends of the orbit are 2 AUs apart.

Such a difference in position enables us to measure the distance of stars up to values of about 100 light-years. (A light-year is equal to 63,225 AUs.) These distances serve as a basis for the estimation of distance of still farther objects by somewhat less reliable methods.

The pictures of stars that the TAU probe will send us

will show them at distances from us five hundred times as great as the extreme width of Earth's orbit. By comparing the distant pictures with those we get from Earth, we will see much larger shifts of parallax and will be able to measure accurately the distances of objects as much as 1.5 million light-years away. Our knowledge of the dimensions of the universe will be enormously sharpened.

However, astronomers will have to wait fifty years after the probe is launched to get the final and best results. What's more, it is not likely that the launching can take place before the year 2000 because we must develop a reliable nuclear-powered engine that can heat and expel the xenon gas. In addition, we must also work out a laser communication system that will reach across the distance of 1,000 AUs. Still, it's rather pleasant that astronomers are thinking of such "far-out" projects.

And just to show it in perspective, even 1,000 AUs is only about one-two hundred seventieth the distance to the nearest star. Think how much more we must do to be able to reach the stars.

The Giveaway Bursts

If there is any such thing as antimatter in the universe, scientists may soon have a way of detecting it.

At one time, scientists thought there had to be antimatter. For every bit of matter created, an equivalent bit of antimatter ought to have been created. The two would be opposites. Wherever matter has a positive electric charge, antimatter has a negative one, and vice versa. Wherever matter has a magnetic field pointing north, antimatter has one pointing south, and vice versa.

If quantities of matter and antimatter encounter each other, they cancel, annihilating each other in an explosion a hundred times as powerful as that produced by a hydrogen bomb with fusing material of the same mass.

Scientists can produce tiny fragments of antimatter in the laboratory, but in the natural world we have about us only matter. The Moon is matter, too; if it weren't, our astronauts would have exploded when they touched it. Mars is matter; if it weren't, the *Viking* probes would have exploded. In fact, we are quite certain the entire solar system is matter.

What about other stars, or other galaxies? Perhaps there are antistars and antigalaxies made up of antimatter, and maybe there are equal quantities of matter and antimatter in the universe, except that they exist apart in different places.

Keeping them apart would be difficult, though. There are clouds of dust and gas, here and there, and they are bound to collide and interact on occasion. If a matter cloud encountered an antimatter cloud, there would be bursts of energetic gamma rays of a particular type, but no such bursts have ever been seen.

In fact, scientists rather reluctantly have concluded that the universe is just about all matter and have worked out theories to explain how, in the original creation, a slight excess of matter over antimatter was produced: in the ratio of a billion to one. Out of that slight excess, the universe as we know it was formed.

But are we sure? Might there not yet be antigalaxies somewhere among the hundred billion galaxies or so that exist, or an occasional antistar at least? How else could we tell? Is there anything that reaches us from distant stars and galaxies that we can study that might give us a clue?

Cosmic ray particles reach us from every direction. They are almost entirely matter, with only a tiny fraction of antimatter, but they don't help us. Cosmic ray particles carry an electric charge and therefore follow curved paths through space. Even if we detected a burst of antimatter particles in the cosmic rays we couldn't tell where they came from. We need to study uncharged particles that travel in straight lines so that we can identify their sources.

There are three types of uncharged particles that reach us from outer space. First are *photons*, which carry the energy in ordinary light, as well as radio waves, X rays, gamma rays, and so on. They reach us in enormous quantities from every star and galaxy, but they are useless. There are no such things as "antiphotons." Matter and antimatter alike emit photons. This means we can never identify a quiet antistar or antigalaxy by simply studying the kind of light it sends us.

A second type of uncharged particle is the *graviton*. Gravitons reach us in vast quantities, too, from every star and galaxy, but they carry so little energy we have so far been unable to detect them. Even if we could detect gravitons, there are probably no such things as "antigravitons," so they wouldn't help us detect antistars either.

That leaves a third type of uncharged particle: the *neutrino*. Neutrinos are tiny subatomic particles with no mass and with no charge; they have almost no interaction with matter. But there is such a thing as an antineutrino as well. Stars and galaxies made up of matter give off quantities of neutrinos; antistars and antigalaxies give off quantities of antineutrinos. Unfortunately, neutrinos and antineutrinos are so difficult to detect that those reaching us from stars and galaxies usually pass us by.

However, every once in a while a supernova explodes and, in its first fury, emits an enormous burst of neutrinos, if it is composed of matter—or an enormous burst of antineutrinos, if it is composed of antimatter. The supernova that was seen exploding in the Large Magellanic Cloud in 1987 sent out an enormous burst of trillions upon trillions upon trillions of particles, and nineteen of them were detected on Earth. This was the first time any such particles were detected from beyond our solar system. They were neutrinos, so the Magellanic supernova seems to be composed of matter.

However, plans are afoot to devise more powerful and delicate neutrino detectors. The time may come when bursts from supernovas will routinely be received and analyzed. (There may be ten supernovas a year in our galaxy alone and equal numbers in other galaxies nearby.)

It may be—in fact, it probably will be—that all the bursts will be of neutrinos. Still, if once, just once, a burst of antineutrinos is detected, we will know we have found an antistar (existing, perhaps, in an antigalaxy), and this may help us reevaluate our notions of the nature of the universe and possibly of its birth and death.

The Neutron Surprise

Part of the excitement of science is that even well-known phenomena sometimes turn up surprises. For instance, a subatomic particle called the neutron has been known to scientists for nearly sixty years, and it has been thoroughly studied. Surely there is nothing new we can find out about it. But there is! In recent months, scientists have had to revise their views about how long a neutron may be able to exist on its own.

The neutron is one of two types of particles present in atomic nuclei. The other particle is the proton. When neutrons are associated with protons in such nuclei, they are stable. They can last indefinitely, as long as the universe lasts.

On the other hand, if a neutron exists on its own, outside the nucleus, it is not stable. Sooner or later, it breaks down into a proton, an electron, and an antineutrino. It is not possible to tell just how long an individual neutron may endure before breaking down. It might be a second, it might be a day—it is a matter of chance.

However, if a large number of neutrons are considered, it is possible to determine just how long it will be before half of them break down. That is called the *half-life*. About 1950, the half-life of the neutron was worked out to be 12.5 minutes. That means if you start with a trillion neutrons, half of them will break down in 12.5 minutes, half of those that survive will break down in another 12.5 minutes, and so on, until eventually they're all gone.

There are many other types of unstable subatomic particles, but the neutron is exceptional. Other unstable particles last only a millionth of a second or less before breaking down. Only the neutron lasts for such a long time as 12.5 minutes.

To scientists, this is an inconvenience. If a particle breaks down in a tiny fraction of a second, it hardly has time to move before breaking down. No matter how fast it is going scientists can trace its movements and time its breakdown. The neutron, however, is usually moving very rapidly as it shoots out of a nucleus, and it will travel many miles before breaking down. Scientists can only watch it for a small portion of its path and must calculate the half-life from the few breakdowns they manage to catch.

Besides, the neutron has no electric charge and scientists can only follow moving particles that have an electric charge. The way they can tell that a neutron exists is by observing how it knocks electrically charged electrons out of the atoms it passes through. It can then be judged how neutrons are breaking down by noting the decline in the number of electrons they produce. But the electrons come off at varying speeds and the ones that are moving very slowly or very quickly may be missed.

Recently, scientists have developed methods whereby neutrons are slowed and trapped inside a magnetic field. They can then be watched at leisure, so to speak, and the breakdowns can be followed more accurately.

This is where the surprise comes in. It seems that the half-life of neutrons is not 12.5 minutes but only about 10.1 minutes. The neutron breaks down about 19 percent faster than was thought.

Does it matter? Is there anything to this besides changing a figure in the textbooks? Yes, as a matter of fact, it does matter, for this tells us something about how the universe began.

Currently, the universe is thought to have started with a big bang. It began with a tiny particle containing all the mass of the universe at a temperature that was enormously high. The particle expanded in a tremendous explosion and the temperature dropped. In the space of a few seconds, the temperature dropped to the point where protons and neutrons formed, and in a few more minutes, it dropped to the point where the protons and neutrons could combine to form atomic nuclei.

A proton by itself is a hydrogen nucleus, but if two

protons and two neutrons combine, they form a helium nucleus. Only hydrogen and helium formed after the big bang. More complicated atoms formed later at the center of stars but only in very small quantities. The universe is still 99 percent hydrogen and helium today.

Of course, neutrons began breaking down after they had formed, so the amount of helium that was formed depended on how long the neutrons remained intact. Assuming a half-life of 12.5 minutes, astronomers calculated how much helium should now exist in the universe. They then studied the helium content of hot and glowing clouds of matter in space. From these and from other data, it seemed that the amount of helium actually present in the universe was less than what had been calculated from the big bang theory. That appeared to be a major flaw in the theory.

However, if the new shorter half-life of the neutron is taken into account, the quantity of helium that would have formed, according to theory, matches the quantity of helium observed—and the big bang theory is supported.

The Invisible Dust Clouds

Not everything in the universe can be seen, so astronomers welcome anything that will make the invisible visible. In February 1987 a supernova, 150,000 light-years away, lit up all of space between itself and our instruments and gave us some interesting information.

To be visible, either to our eyes or to special instruments, an object has to emit radiation. Stars do, for instance, as do distant objects made up of stars, such as galaxies and quasars. Even dust clouds can be visible if they happen to contain stars. Starlight is reflected and scattered by the surrounding dust, giving us useful information.

There are dust clouds in space, however, that are nowhere near stars and are therefore cold and dark. Occasionally we see nearby *dark nebulae* of this sort because they block the stars behind them. They then appear as dark shapes within which no stars appear but that are outlined on all sides by a starry blaze. Other dark clouds in our galaxy may be too thin to be made out in this way or too far away to show up easily in this fashion. These dark clouds of dust and gas are of great interest to astronomers.

For one thing, they are the raw material from which new stars are born. Occasionally such clouds condense and grow hotter until they ignite in nuclear fire and become a young star. Nearly 5 billion years ago, it was such a condensing cloud that formed our own solar system. And the process has been continuing ever since. We are actually watching this sort of thing happening in some nearby clouds such as the Orion Nebula, which now shines brightly because of the young stars that have already formed in it.

Second, in some dark clouds in which no stars are yet forming, atoms cling to each other, forming dozens of different combinations. Every different combination emits its own unique radio waves by which we are able to identify it. Some of them can give us an insight into how complex molecules might be built up from their atoms and thus help us speculate on how life might have come into being on Earth. To make these observations we need clouds that are near enough and dense enough.

There must be many dark clouds in our own galaxy, however, that are simply too far away or too thin (or both) to be observed and studied, unless somehow we could turn a very bright searchlight on them. Such a searchlight appeared, at least in one particular direction, when the supernova in the Large Magellanic Cloud blazed out.

The light, as it reached us across a gap of 150,000 light-years (nearly a billion billion miles), passed through thin clouds of matter lying within the Large Magellanic Cloud, then through other clouds that lay between that cloud and our galaxy, and finally through still other clouds that lay within our galaxy. As the light of the supernova passed through each of these clouds, some of that light was absorbed and from the nature of the absorption, astronomers could deduce a great many things.

For instance, astronomers calculated that as the light of the supernova journeyed toward us over the light-years, it passed through twelve clouds in the Large Magellanic Cloud, then twenty-two clouds in the intergalactic space between the cloud and our galaxy, then six more in our galaxy. That's forty clouds altogether that had been invisible to us until the supernova exploded.

From the nature of the absorbed light, astronomers can further deduce that the Milky Way galaxy (at least that portion of it through which the light traveled) is a very dusty place, and that the clouds consist of gas as well as dust. The Large Magellanic Cloud is somewhat less dusty than our galaxy is (but then the cloud has only one-tenth as many stars as our galaxy has, and they are spread out at greater distances from each other). The intergalactic space between our galaxy and the cloud seems not to be dusty at all, so that the clouds there consist largely or entirely of gas.

So far, astronomers have known of clouds only in the various galaxies, particularly our own, of course, though dark areas could be seen in others as well. They have known virtually nothing of clouds of matter between the galaxies.

The guess is that the Large Magellanic Cloud, being a

small galaxy that is nearer our own Milky Way than any other galaxy, exerts a considerable gravitational pull on us, and we on it, of course. This would be especially so if the cloud had been closer to us in the past and had, indeed, skimmed our edge. In that case the mutual gravitational attraction, although not strong enough to disturb individual stars or dust particles very much, would have been able to pull out quantities of individual atoms and set up a string of gas clouds between the two galaxies.

Parts of those intergalactic gas clouds are surprisingly hot, and some also contain the ordinarily rare element lithium. Both factors cry out for explanation, and astronomers, like the rest of us, love a mystery.

The Weakest Wave

Is it possible to design an instrument that can detect gravity waves?

According to Albert Einstein's theories, waves of gravitation should exist. But if they do, they are so faint that scientists have never been able to detect them. They are still trying, though, and they may succeed.

How do we know they're there if we can't detect them? Einstein worked out the general theory of relativity in 1916 and showed that the presence of matter distorted space, resulting in the gravitational force. Every time matter is redistributed in space, the nature of the distortion changes and this produces a disturbance, a *gravitational wave*, that spreads outward in all directions at the speed of light.

Astronomers are quite certain by now that the theory of general relativity is correct, so that these gravitational waves must exist. The Earth must give them off as it turns around the Sun, for instance. The Earth loses energy in this way and therefore is spiraling gradually inward toward the Sun.

In that case, why don't we detect these gravitational waves? The answer is that gravity is by far the weakest force we know. The electromagnetic force that holds atoms together is a thousand trillion trillion trillion times as intense as gravitation. The only reason we're so aware of gravitation is that the Earth is an enormous body and the gravitational pull of its myriads of particles adds up to something noticeable.

Gravitational waves are, therefore, the weakest and shallowest waves that exist, and they simply don't produce any effects we can detect. The amount of energy Earth loses by gravitational waves is so tiny that even in its billions of years of existence, Earth has spiraled inward toward the Sun only a trifling distance.

Naturally, more energetic redistributions of mass will produce stronger gravitational waves. A really massive redistribution, such as the collapse of a star into a black hole or a collision of two stars, might produce gravitational waves just strong enough to detect. If so, a detecting instrument can give us information about the really great catastrophes that may be taking place here and there in the universe, information we might get in no other way.

Back in the 1960s, a scientist named Joseph Weber at the University of Maryland tried to detect gravitational waves. He used large aluminum cylinders. If a gravitational wave swept over it, a cylinder would compress and expand by a distance of about a ten-millionth the width of an atom.

Nevertheless, the strongest gravitational waves might produce a compression just large enough to be detected.

To make sure whatever was detected actually was a gravitational wave, Weber made use of two cylinders, one in Maryland and one in Illinois. A gravitational wave would be so long and shallow it would encompass the entire Earth and should affect both cylinders simultaneously. Weber thought he detected the waves and for a while there was considerable excitement. However, others could not repeat the experiment, and the feeling was that although Weber did important work, his instruments simply weren't quite delicate enough for the job.

Scientists refuse to give up. The desires to have another demonstration of the truth of general relativity and to be able to detect whispers of great events in the distance keep them working on new "gravitational telescopes."

One promising design for such a "telescope" is being considered at the University of Glasgow in Scotland by a team under Jim Hough. The instrument would consist of two evacuated tubes (tubes from which all the air has been removed, creating a vacuum) at right angles. In each tube, a beam of laser light would be reflected back and forth a thousand times or so. If the tubes are completely undisturbed, the light wave will remain in perfect step.

If a gravitational wave swept over the tubes, however, one tube would be compressed a very tiny bit more than the other, and this would throw the laser beams out of step. This could then be detected, making it possible not only to spot a gravitational wave but to estimate its energy content and obtain some information about what might have produced it.

Right now, the Glasgow people are working with tubes that are each ten meters (about thirty-three feet) long, just to test the workings of laser beams. Matters look promising, but what they will need eventually, if they are to have a chance of detecting gravitational waves, are tubes about a kilometer (five-eighths of a mile) long. Such an instrument would cost about $25 million to produce.

What's more, to do it properly, there should be four such instruments distributed over the whole world, so that all four would be affected almost simultaneously to make certain a gravitational wave and not something else was affecting them. There would be tiny differences in the time of detection, because it would take the wave about one-twenty-third of a second, moving at the speed of light, to pass from one end of the Earth to the other. By working with such tiny time discrepancies, it might be possible to locate the direction from which the waves are coming.

The researchers now are in the process of trying to raise the money for such a large undertaking.

The Relativity Test

Einstein's theory of relativity rests on a certain assumption, and for eighty-four years, scientists have been testing the assumption. It has passed the test every time. Still, they keep testing, because even if the assumption is only slightly off, this might open the way to a new theory that would be even broader, more useful, and more nearly correct than relativity. In early 1989, the theory was subjected to still another test and Einstein's assumption passed again.

The assumption is this: The speed of light is always the same, regardless of the speed of the light source.

This isn't the way ordinary moving objects seem to behave. If you throw a ball from a moving train in the direction the train is moving, the ball moves faster through the air. If you throw it in the opposite direction from which the train is moving, the ball moves slower. The speed of the source adds to the speed of the ball if both are moving in the same direction. The speed of the source subtracts from the speed of the ball if both move in opposite directions.

It seemed to Einstein, however, that this would not be true of light or for anything else that moved at the speed of light. Speeds would, in that case, not add together or subtract but always stay the same.

If this is true, it would mean that the faster an object moves, the less it is affected by the speed of the source until at the speed of light it isn't affected at all. Einstein worked out an equation to show how the speed of a source would add to or subtract from the speed of an object depending on how, and how fast, the two were moving in relation to each other.

He also deduced that with increasing speed, objects would become shorter in the direction of travel; that they would grow more massive; that they would experience time more slowly; that nothing with mass (for example, we and our spaceships) could ever go faster than the speed of light.

All this seems contrary to "common sense" and hard to believe because we are surrounded by objects moving at very much less than the speed of light, so that we are used to speeds adding and subtracting by simple arithmetic. However, as scientists began to study objects that moved very rapidly, such as speeding subatomic particles, they found that Einstein's deductions were correct in every particular. Atom-smashing machines wouldn't work the way they do if Einstein's deductions weren't correct; nor would nuclear bombs explode.

Naturally, if the deductions are correct, we have to suppose that the original assumption must be correct, too.

You don't get correct deductions from false assumptions. But perhaps the assumption, and the deductions that flow from it, is only *nearly* correct. This would, as I've said, place us on the track of something even better than relativity. So scientists keep testing the assumption.

Well, in February 1987, light from a star that exploded into a supernova, about 160,000 light-years away from us, finally reached us. *Neutrinos* (massless subatomic particles that travel at the speed of light) also reached us from the supernova. Einstein's assumption holds for neutrinos, too. They also travel at the same speed regardless of the speed of the source.

Every bit of an exploding star emits neutrinos in all directions. Some neutrinos are radiated in our direction from every bit of the explosion, and we can detect them—not many, for neutrinos are fearfully hard to detect, but some.

The bits of the exploding star move at sizable fractions of the speed of light. Some bits move away from us rapidly. Some bits move toward us just as rapidly. Some bits move crossways, and every direction in between, just as rapidly. If speeds only added and subtracted, neutrinos from parts of the explosion moving away from us would travel more slowly in our direction and arrive much later than neutrinos from parts of the explosion moving toward us. On the other hand, if neutrino speeds were unaffected by the speed of the source, all the neutrinos ought to reach us at exactly the same time, no matter what part of the explosion they came from.

Astronomers detected just nineteen of the neutrinos, which all reached the detecting instruments in a burst of 12 seconds: none earlier, none later. The neutrinos had been traveling for 160,000 years (particles traveling at light speed take 1 year of time to travel 1 light-year of distance). In each year, there are 31.55 million seconds. This means that the neutrinos had been traveling for 5 trillion seconds, and yet the spread among them was only 12 seconds.

Kenneth Brecher and Joao L. Yun of Boston University used the data obtained from the neutrino observations and showed that they indicated that Einstein's assumption is correct to better than 1 part in 100 billion. This means that the speed of light (186,282 miles per second) might vary, at most, one-tenth of an inch per second, either more or less.

That is a tighter test than any that Einstein's theory has ever received in all the eighty-four years since the assumption was first advanced, so the theory of relativity looks better than ever.

Neutrinos from Afar

The new supernova in the Large Magellanic Cloud is the closest supernova in nearly four hundred years. It now seems to have produced a bit of headline news on the subject of neutrinos.

Neutrinos are the tiny particles without mass or electric charge that travel at the speed of light and pass through matter as if it weren't there. Neutrinos will sweep right through the Earth from end to end without being stopped or even slowed—well, almost. One neutrino out of many trillions will be stopped.

Physicists have devised setups that will stop and detect the occasional neutrino. In this way, neutrinos, which were predicted from theoretical calculations in 1931, were finally detected in 1956 in fission reactors, which produce large quantities of them.

In the last few years, physicists have set up "neutrino detectors" deep in the Earth, in order to record neutrinos produced by the Sun. The detectors must be buried deep so that no other particles will penetrate and confuse matters. Neutrinos from the Sun were detected but in smaller quantities than expected (something that is still a mystery).

However, neutrinos have never been detected from any source other than the Earth and the Sun. Neutrinos from other stars are so thinned out by distance that too few reach us to give us a decent chance at detection—until recently.

The new supernova, at the very start of its explosion, apparently sent out a vast flood of neutrinos. Thanks to the supernova's closeness to us, enough neutrinos reached us to be spotted by a neutrino-detection device deep under Mont Blanc in the Alps. This device is operated by Italian and Soviet physicists.

To me, at least, the discovery came as no surprise. Back in 1961, I was corresponding with a young physicist named Hong Yee Chiu, who had been at Cornell and then joined the Institute for Advanced Study at Princeton.

He was interested in supernovas, and he did his best to calculate what nuclear reactions would take place at the center of a giant star as it grew older and older and hotter and hotter. It seemed to Hong Yee Chiu that eventually the star's center would reach a temperature of 6 billion degrees C (four hundred times as hot as the center of our Sun). At this enormous temperature, particle interactions that produce neutrinos should, he decided, become very prominent. Enormous quantities (a quadrillion times as many as the Sun produces) of neutrinos would be formed. Whereas other particles would be more or less trapped in the star's center and would make their way to the outer regions very slowly, the neutrinos, unaffected by matter, would leave the center at the speed of light, carrying energy with them.

The center of the star, losing these neutrinos and their energy, would cool off with catastrophic quickness. The center would no longer be hot enough to support the weight of the upper layers and the star would collapse, producing a supernova explosion and leaving behind, finally, a neutron star or a black hole (though those terms weren't yet in common usage in 1961).

"Therefore," said Hong Yee Chiu, in one of his papers, "the establishment of a neutrino monitor station in terrestrial or spatial laboratories may help us predict forthcoming supernovas."

I don't know that Hong Yee Chiu's predictions received much attention a quarter-century ago, but they impressed me tremendously. I described his conclusions in an article titled "Hot Stuff" in July 1962. Now that his theory may have been borne out, I would like to see him get the recognition.

My reasons are partly selfish, I suppose. Hong Yee Chiu had been working on particle physics at Cornell and then switched to the field of astrophysics and became interested in supernovas. The reason for his switch was given in a letter to me, which I quote: "I switched from the field of elementary particle physics to astrophysics then, right after I got my degree. Your article (of October 1959) initiated my interest in the field of supernovas."

My article was titled "The Height of Up." It had nothing to do with supernovas, but it did ask how high a temperature could be reached in our present-day universe. I worked out an answer for myself in a crude sort of way, but Hong Yee Chiu, reading the article, thought he'd try his (more expert) hand at it. It seemed to him that temperatures were highest at the center of stars, particularly giant stars, and most particularly giant stars that had heated to the point of explosion. He thus found himself working with supernovas.

I feel great about this. Though educated to be a scientist, I chose writing as my profession. I am never likely, therefore, to make any scientific discoveries myself, but it pleases me every bit as much when my speculations inspire others to make them.

The White Dwarf Clock

How old is the universe? The question haunts astronomers, and an answer based on a new type of "clock" has now been offered.

For sixty years, clues to the universe's age have been sought in the rate at which the universe is expanding. Once the rate is known, astronomers can judge how long a time was required for the universe to expand from a tiny point to its present size.

Unfortunately, the exact rate of expansion is difficult to determine, meaning that estimates of the age of the universe can only be crude. As a result, they range from 10 billion years to 20 billion years. Many people arbitrarily choose 15 billion years as the age of the universe because it is between the two extremes.

Another way is to determine the age of the oldest stars by studying their chemical composition and judging how long it would take some long-lived atoms to break down to the levels found in those stars. This method seems to show the age of the universe as 10 billion years.

Now a third method for determining the age has been worked out. It involves "white dwarfs."

Ordinary stars like our Sun eventually run low on the nuclear fuel that keeps them shining, usually after billions of years. When this happens, stars become red giants. This occurs when stars expand and their surface cools down. Then, without enough fuel to keep its structure expanded, the star collapses into a white-hot object with the mass of a star but a size no larger than the Earth's. It is then a white dwarf.

Some particularly large stars collapse even more drastically, becoming tiny neutron stars, only eight miles across or so, or even into still smaller black holes, which are believed to be invisible collapsed stars so condensed that neither light nor matter can escape from their gravitational fields. But generally, most stars the size of our Sun or smaller become white dwarfs.

In white dwarfs there are no nuclear reactions giving rise to heat and light. White dwarfs have only the energy they started with as a result of their collapse. As time goes by, therefore, they radiate away this fixed amount of energy and gradually grow dim.

Of course, white dwarfs are so small that even when they are brand-new and very hot and bright, the total amount of light they give off is small compared with that of our large Sun. This means that although there are at least a billion white dwarfs in our galaxy alone, we can only see those that are fairly close to us.

Even so, this gives us quite a number to work with. Some white dwarfs seem dim to us because they *are* dim. Some, however, seem dim only because they happen to be farther away than most others. If we measure the distance of various white dwarfs, we can make allowance for this and determine what the brightness would be if all the white dwarfs were the same distance from us. This is their *luminosity.*

The older the white dwarf, the lower the luminosity. Would some of them be so old that they had lost all their energy and were simply dark cinders invisible to the eye? Apparently not. White dwarfs have so much mass and cool off so slowly that the time they would take to turn into "black dwarfs" is much longer than any estimate of the universe's lifetime. Therefore, any white dwarf that ever formed should still be luminous to some degree.

This means that though the very oldest white dwarfs would be the dimmest, they would still be shining and visible. At the University of Texas, the astronomer Donald E. Winget and his colleagues have calculated the luminosities of many white dwarfs. Very luminous ones are rare because they include only those that formed very recently. Less luminous ones are more common because they have formed over longer periods of time.

However, it becomes almost impossible to find white dwarfs below a certain luminosity, even though at this level they should still be easily visible. Apparently, the dimmest white dwarfs formed when the universe was very young. Before that, stars hadn't lived long enough yet to produce white dwarfs.

Calculating the time it might have taken for the dimmest white dwarfs to grow so dim, and adding an extra billion years for stars to shine before they reached the white dwarf stage, the universe would seem, again, to be 10 billion years old. Thus, a figure of 10 billion years can be reached by three widely different methods: rate of expansion of the universe, rate of breakdown of radioactive elements, and rate of dimming of white dwarfs.

Incidentally, for those who are fascinated by numbers,

10 billion is a 1 followed by 10 zeros, which means it is the product of 10 10s. Isn't that a neatly round number for the age of the universe?

The Gamma Giveaway

The most peculiar creature in the astronomers' zoo is the "black hole" and astronomers are laboring to determine whether it really exists. Recently, some evidence has appeared that seems to indicate that it does.

It is possible, you see, for a star to collapse so that its atomic nuclei come into contact and it becomes a *neutron star.* When this happens, a star the size of the Sun collapses into a small sphere only eight miles across that still contains its entire mass. Its gravitational field then becomes monstrously intense: a teaspoon of its matter weighs a million tons. Light itself can barely escape from it.

Neutron stars were discovered only in 1969, and astronomers are certain they exist. The tiny stars turn rapidly, in periods that are anywhere from one turn per second to nearly one thousand turns per second, and we can detect the radio pulses sent out with each turn. There are a few neutron stars we can actually see, as they send out pulses of light and blink rapidly on and off.

If a neutron star is too large, however, its gravity causes the atomic nuclei themselves to collapse. The star then shrinks down to virtually nothing, and its gravitational intensity grows larger without limit. Things can fall into such a collapsed star, but nothing can battle its gravitation and emerge, so it's like an infinitely deep hole in space. Not even light can get out, so it is a *black* hole.

But do black holes actually exist? The centers of many galaxies give off energetic radiation in huge quantities, and the easiest way of explaining this is to suppose there are enormous black holes there. Even our own galaxy seems to have a large black hole at its center. However, the evidence is indirect and not entirely convincing.

The nearest object to us that may be a black hole is something called Cygnus X-1, which is a source of X rays. Near the site of Cygnus X-1, a giant star that is about thirty times as massive as our Sun is visible. It seems to be moving in space as if it and Cygnus X-1 were revolving about each other. From the nature of the star's motion it would seem that Cygnus X-1 is from five to eight times the mass of the Sun. Yet at the site of Cygnus X-1, nothing can be seen. All we can detect are X rays.

We might suppose Cygnus X-1 is a neutron star, too small to see, with its radio pulses not aimed in our direction. This can't be, though, because a neutron star can't be more than three and one-third times as massive as our Sun. If it is any more massive, its gravitational force would be large enough to make it collapse into a black hole. Cygnus X-1, therefore, must be a black hole.

That seems to be pretty conclusive, but it depends on how far from us that circling pair of objects might be. If they were closer than we think, then the giant star and the X-ray source would be nearer each other than we think. Their motion then could be brought about by smaller masses.

We think that Cygnus X-1 is 10,000 light-years away from us, but what if it were only 3,000 light-years away? In that case, the giant star would be only ten times the mass of our Sun, and Cygnus X-1 might be only two times the mass of our Sun. It could then be a neutron star and not a black hole.

The X rays that emerge from Cygnus X-1 would be expected if it were a black hole. Matter from its companion star would be pulled into the black hole and as the matter spiraled down it would emit X rays. However, X rays also would be emitted if the object were a neutron star. But a neutron star would not emit gamma rays.

A satellite called "High Energy Astrophysics Observatory 3," which has been detecting X rays coming from Cygnus X-1, has also been detecting gamma rays. Gamma rays are like X rays but are made up of tiny waves even shorter than those of X rays. Gamma rays, therefore, are the more energetic of the two. The gamma rays observed are up to one thousand times as energetic as the X rays that emerge from Cygnus X-1.

Neutron stars have gravitational fields strong enough to cause matter to spiral in tight turns that automatically emit X rays, but the fields cannot force the matter to spiral so tightly that it will emit gamma rays. Black holes, however, with still greater gravitational intensities, can produce gamma rays.

Early in 1988, astronomers at the Jet Propulsion Laboratory in Pasadena, California, reported a new way of looking at the matter. The gamma rays seem to be coming from a small region only about three hundred miles across where there must be gas at a temperature of several billion degrees C. At such a temperature this gas produces electron-positron pairs that annihilate each other and produce gamma rays.

Neutron stars just don't have the energy to do this, but black holes do. This new information, then, confirms that Cygnus X-1 is a black hole.

The Force that Can Swallow a Star

It's the force of gravitation that holds the universe together. Every bit of matter, no matter how small or how large, produces a gravitational pull. The larger the bit of matter, the more mass it has and the more intense the pull. As the mass is concentrated into a smaller and smaller volume, the pull grows still more intense.

In order to escape from the gravitational pull of a large object, a small object must leave it at more than a certain speed called an *escape velocity.* The stronger the gravitational pull, the greater the speed required to escape. There is no limit to the strength of the gravitational pull, but there is a limit to speed. Nothing can go faster than the speed of light, which is 186,282 miles per second. If an object is massive enough and the mass is concentrated enough, even light doesn't move fast enough to escape. Therefore, nothing else does either.

A massive, concentrated object from which nothing can escape is like an infinitely deep hole in space. Anything can fall in, but nothing can come out. Since even light can't come out, it is totally black, explaining why it is called a "black hole."

Astronomers believe black holes exist, but how can they detect something that is totally black and gives out no light? Well, suppose the black hole happens to be located near a quantity of matter; suppose, for instance, that it and an ordinary star are revolving about each other. If the star is near enough to the black hole some of the star's matter may be pulled into the neighborhood of the black hole.

Such matter circles the black hole just as a planet would, spiraling slowly inward until, little by little, it falls into the black hole. We know from experiments here on Earth that matter circling in a gravitational field in that fashion gives off energy. The energy given off by the matter circling a black hole is huge, and it emerges in the form of a stream of X rays. We might consider this (if we want to be dramatic) the shriek of dying matter.

There are places in the sky where astronomers find X-ray sources that don't seem to be visible to the eye. Where the X-ray sources possess certain characteristics, it seems reasonable to suppose that they represent a black hole that is in the process of swallowing matter. Fortunately, none of these is close to us. Even the nearest must be at least 5,000 light-years away, millions of times as far from us as the distant planet Pluto.

When a black hole swallows matter, it grows larger. Naturally, such growth occurs most readily where there is a great deal of matter in the vicinity for the black hole to swallow. Astronomers find X-ray activity in the center of globular clusters, for instance, which are closely spaced crowds of up to hundreds of thousands of stars. There are several hundred such clusters in our galaxy.

Galaxies are made up of billions of stars, sometimes trillions of them, and the crowds of stars at the center of galaxies are far more numerous and far more closely spaced than in globular clusters. And, indeed, astronomers find that the centers of many galaxies are "active," producing floods of X rays and other radiation. They suspect that the largest black holes of all are to be found there.

There are some galaxies in which the centers are particularly active. These are called Seyfert galaxies, after the astronomer who first described one. Seyfert galaxies must contain black holes that are truly monsters, with masses equal to millions or even tens of millions of ordinary stars.

The larger a black hole, the greater the mass it can swallow. It would seem that a black hole that is large enough might swallow a whole star at a gulp. This may be true of the black holes at the center of Seyfert galaxies, for instance.

Two astronomers from Ohio State University have been watching a Seyfert galaxy called NGC 5548, which is thought to have a black hole at its center with a mass equal to 30 million stars. Recently, they detected a sudden burst of radiation from it. From the size and nature of the radiation, they suspect that happened because the black hole drew in a star some four-fifths the size of our Sun and swallowed it whole.

Is our own Sun in danger of such an end? Not really. The nearest black hole that might conceivably swallow it is at the center of our own galaxy, and this is 30,000 light-years away. Our Sun and its planets circle that center without ever getting much closer to it than we are right now.

Of course, long before we are close enough to be swallowed by a black hole, the flood of X rays and other radiation arising from the matter that is continually falling into it would make life impossible on Earth. Fortunately, even a near approach is not likely.

The Cluster Yardstick

Astronomers may have a new way of measuring the distances of some galaxies.

This is badly needed. Without an accurate knowledge of galactic distances, we can't tell for sure how far apart the galaxies are and how quickly the universe is expanding. This means we can't be very precise about how old the universe is. Some astronomers say the universe is 10 billion years old, and some say 20 billion years old: a substantial difference. This makes us uncertain of other key features of the universe and of what its ultimate destiny may be.

The best way of determining a galaxy's distance from Earth is to detect a type of star in it called a Cepheid variable, whose light grows brighter and dimmer over a certain period. By measuring the period, we can calculate how luminous it really is: the longer the period, the greater the luminosity. The dimmer the star appears, the farther away it is; from its apparent dimness, we can calculate just how far away it is.

The Cepheids allow us to calculate the distance of the thirty nearest galaxies quite accurately. The Magellanic Clouds are 150,000 light-years away, for instance, and the Andromeda galaxy is 2.2 million light-years away. Beyond those thirty, though, the Cepheid variables become too dim to see at all. Thirty isn't many among the billions of galaxies.

There are ways of determining the distance of galaxies that are farther off, but they're not as good as the Cepheid method. There are giant stars, for instance, that are much brighter than Cepheids and can be seen at six times the distance. A really bright star is about a million times as bright as the Sun, and from the degree of its dimness we can tell how far away it is. But determining whether a star is a million times as bright as the Sun is a matter of guesswork. Also, not every galaxy contains these giant stars.

We also can detect supernovas occasionally in some galaxies, even those that are very far away, because a supernova shines with a light that may be tens of billions of times as bright as that of our Sun. But again we have to guess at how bright the supernova actually is, and they appear only in scattered galaxies here and there.

Finally, we are reduced to judging the distance of a galaxy by its total brightness because it is too far away to see any ordinary stars in it, even the brightest, and supernovas are rare. This would be a good method of determining distances if all galaxies were of the same size, but some are a million times as large as others. So judging distance by overall brightness is very risky.

This brings us to *globular clusters*. There exist, apparently in every galaxy, groups of stars that are packed tightly together in a spherical shape. Such globular clusters have various sizes. The smallest may consist of a few tens of thousands of stars, and the largest may consist of a million stars.

Our own Milky Way galaxy contains about two hundred of them that we can see and perhaps one hundred more hidden by dust clouds. The Andromeda galaxy contains a similar number of these globular clusters, and they have been detected in other nearby galaxies, too.

It may be that globular clusters can only reach a certain maximum size. Perhaps if they were any bigger the outermost stars would not be held firmly enough by the gravitational pull of the inner ones and would eventually separate from the cluster. If this were so, then the brightest globular clusters would always have the same total brightness.

William Harris, an astronomer at McMaster University in Hamilton, Canada, has recently made a careful study of the globular clusters in nearby galaxies whose distance is known. He determined the apparent brightness of the globular clusters, and from their distance he calculated how luminous they really are in comparison to our Sun, for instance. He reported that, indeed, the brightest globular clusters, in any galaxy, shone with about the same luminosity.

This means that if we can make out globular clusters in some galaxy whose distance we're not certain of and determine the brightness of the brightest of them, we can compare this with the luminosity they really have and calculate the distance.

In a way, it might not seem this gives us anything new. A bright globular cluster is about as bright as the brightest supergiant single stars, and we use those superbright stars already to determine galactic distances. The advantage of the globular clusters, however, is that they probably occur in every galaxy, and supergiants only in some. Second, if Harris is correct, the brightness of the brightest globular clusters can be relied on more safely than the brightness of the brightest individual supergiants.

The globular cluster method may, therefore, give us *reliable* distances for the nearest six thousand galaxies. This will be an important improvement over the present situation, but there are still billions of other galaxies. We must continue to look for additional yardsticks.

Gravity Plays Tricks

Something that seemed very exciting at the beginning of 1987 now appears to have been an optical illusion. Some luminous semicircular arcs of light that were found to circle distant galaxies seemed to be the longest objects ever seen and had astronomers guessing. The solution to the mystery involves the force of gravity.

Back in 1916, Albert Einstein, in his general theory of relativity, showed that light rays bent slightly when they passed some massive object. Suppose then that light rays from some distant object passed on all sides of a massive object on their way to us. The rays would be bent inward on all sides and might come to a focus at our eyes. In that case, the distant object would seem to be enlarged just as if it had passed through the lens of a magnifying glass. For this reason, this action of gravity is called a *gravitational lens.* Einstein predicted that these lenses ought to exist.

However, light is bent so slightly even by massive objects that in order for the rays to come into focus, they must travel a long distance. This means that the object that does the focusing must be very far away and the object that sends out the light beams must be much farther away still.

In 1916 and for nearly half a century later, astronomers didn't know of any objects that were so far away that a gravitational lens was possible. The phenomenon was therefore considered theoretically possible but unlikely to be encountered in actual fact.

However, in the early 1960s, *quasars* were discovered.

These were galaxies that had very active and luminous centers and were very far away. Even the nearest quasar was 1 billion light-years from us, much more distant than anything previously known, and some were 10 billion light-years away. They are so far away that all we can make out is the tiny luminous center, which looks like a dim, very distant star.

Suppose there happened to be an ordinary galaxy between the quasar and us. The ordinary galaxy might be far enough away to be only dimly visible, if at all, but it would be considerably closer than the quasar. The quasar's light, on its way to us, would pass by the ordinary galaxy on different sides and reach us out of focus. In that case, we are likely to see two images of the quasar, one on each side of the ordinary galaxy. This would be the effect of a gravitational lens.

In 1979, astronomers at the University of Arizona detected two quasars, very close to each other, that seemed very similar in appearance, in brightness, and in the properties of the light they emitted. Could they be two images of the same galaxy produced by a gravitational lens? If so, there should be an ordinary galaxy present between us and the quasar. It was searched for, and a very dim one was found. The first case of a gravitational lens had been detected.

Naturally, other examples were sought and astronomers now think they may have found as many as seven such distorted quasar images produced by gravitational lenses.

That brings us to the luminous arcs that surround a few distant galaxies and that were found early in 1987 by Vabe Petrosian of Stanford and C. Roger Lynds of the Kitt Peak National Observatory near Tucson, Arizona. These arcs were rich in ultraviolet light and seemed to be smoothly and perfectly circular; one is as much as 325,000 light-years long—much longer than our galaxy.

It occurred to Petrosian and Lynds that if a quasar

were centered exactly behind a galaxy, the image should appear on all sides equally and take on the appearance of a circular arc or part of one. They looked for dim galaxies that might lie between the circular arcs and us, found them in two cases, and announced in November 1987 that the arcs must be produced by a gravitational lens.

These gravitational lenses may do more than offer us a spectacular sight (at least those of us with a large enough telescope). They may help us solve a puzzle. Most astronomers think the universe may stop expanding some distant day in the future and begin to contract. However, the amount of matter we see in the universe is only about 10 percent of the amount needed to produce sufficient gravitational pull to stop the expansion.

Does this mean that the universe may never stop expanding, after all? Or is there some kind of matter present that we can't see and that makes up 90 percent of the universe?

Well, the galaxies that lie between us and the luminous arcs don't seem to have enough matter in them to bend the quasar's light sufficiently to produce the effect. Yet the fact is that they do bend the light. The conclusion is that they are much more massive than they appear to be, and this means that they must possess matter we don't detect in the usual way.

But what can this undetectable matter be? This problem is called the "mystery of the missing mass," and it may be that these luminous arcs will give us hints to a solution of that mystery.

The In-Between Objects and the Missing Mass

Astronomers know about stars and know about planets, but they are now busily engaged in searching for bodies that are in-between, that are too small to be stars and too large to be planets. These are very elusive objects, but their existence—if they exist—could be very important.

The star we know best, of course, is our own Sun, and the largest planet we know is Jupiter. The mass of the Sun, the amount of matter it contains, is about one thousand times that of Jupiter.

The Sun is so massive that the hydrogen atoms at its core are crushed and pressed together at very high temperatures and therefore undergo fusion. This releases vast amounts of energy, which is why the Sun shines and has been shining for billions of years. Jupiter is simply not large enough to have atoms at its core pressed into heat and fusion. It is too cold to shine of itself; we see it only by its reflection of the light of the Sun.

Stars come in all sizes, of course. The larger a star is, the more energy is produced at its core and the brighter and hotter it shines. There are stars up to sixty or so times the mass of the Sun. On the other hand, stars that are smaller than the Sun are dimmer and cooler. Some stars may be only one-tenth the mass of the Sun.

The smallest stars we see are no more than red-hot and are therefore called *red dwarfs*. They are too dim to see at great distances, and we can study them best when they are not many light-years away. The smallest red dwarfs may be only about one hundred times the mass of Jupiter.

If a red dwarf is so small that it simply doesn't produce enough heat to shine with any visible light of its own, then it might be called a *black dwarf.* Jupiter might be viewed as such a black dwarf. So might Earth and all the other planets we know.

But what if objects form in space that are more massive than Jupiter, but less massive than red dwarfs—objects that are, say, ten to eighty times as massive as Jupiter? They aren't massive enough to produce hydrogen fusion and shine like a star, even a dim star. On the other hand, they might be massive enough to bring about some forms of nuclear reaction other than ordinary hydrogen fusion. Or else their gravitational pull might be large enough, all by itself, to heat them to the point that they radiate small amounts of energy.

Such in-between objects might produce a very dim red light. They might also produce quantities of the less energetic infrared light that our eyes can't see but our instruments can detect. Such objects wouldn't be quite black, so they are now called "brown dwarfs." (This is a poor name because they are not actually brown; they should more properly be called "infrared dwarfs.")

Smaller stars are more common than large ones. Large, massive stars are very rare, and even middle-size stars like the Sun are relatively few in number. At least three-quarters of all the stars bright enough to shine are red dwarfs. It follows that the still smaller brown dwarfs ought to be very common indeed.

If so, this is important. Astronomers have known for years that galaxies seem to have considerably more mass than can be accounted for by the stars we see in them. This is called the "mystery of the missing mass." If, however, every galaxy (including our own) has vast crowds of brown dwarfs, which we don't see, this would account for at least part of the missing mass, if not all of it. Confirmation of such additional mass would sharpen our theories about the future development, and ultimate fate, of the universe.

On the other hand, it may be that the mechanics of star formation is such that no objects with sizes of from ten to eighty times the mass of Jupiter are formed. In that case, explaining the mystery of the missing mass would become much harder. It is for this reason that astronomers are searching for any signs that brown dwarfs exist. In recent years, brown dwarfs have been reported, but the reports, till now, have turned out to be premature.

The best places to look, it would seem, would be in relatively nearby areas where active star formation is going on right now, where new young stars might be detected. Among them might be some objects small enough to be brown dwarfs. There is such a region in the constellation of Taurus, and earlier this year, a group of astronomers headed by William Forrest of the University of Rochester in New York reported having detected objects there that could be brown dwarfs.

These objects give off long-wave radiation that instruments pick up, and they do not seem to be circling ordinary stars, so that they are not planets that are reflecting radiation from stars. They are independent objects. Forrest estimates that these brown dwarfs may be from five to fifteen times as massive as Jupiter. What's more, from the fact that he found seven of these objects in a tiny area, he estimates that there may be at least one hundred brown dwarfs in this region for every ordinary star. If this is confirmed, and if it is found to be true in other areas of space as well, the mystery of the missing mass may have been solved.

Galaxies in Collision

If the Earth or the Sun were involved in a major collision with some other astronomical body, it might very likely mean the end of life on our planet. But what if our entire galaxy, containing our Sun and about 200 billion other stars, underwent a major collision?

Such a collision is not likely in the immediate future, but if we wait long enough, it will surely happen. Our Milky Way galaxy is not alone in space, you see. It is part of a cluster of about two dozen galaxies that are, all together, known as the Local Group.

Most of the galaxies of the Local Group are dwarfs, each containing only a few billion stars. An example is the Large Magellanic Cloud, only 150,000 light-years away and containing 10 billion stars. (It is much in the news, because a supernova blazed out in it in February 1987, the closest supernova to Earth in almost four hundred years.)

The galaxies in the Local Group are not all dwarfs, however. One, Andromeda galaxy, a giant even larger than our own, contains perhaps 300 billion stars. It is closer to Earth than any other giant galaxy: about 2.3 million light-years away. This is more than fifteen times the distance of the Large Magellanic Cloud.

These two giant galaxies, our Milky Way and the Andromeda, aren't immovably fixed at this distance. The two galaxies move about a common center of gravity. What's more, the orbits are quite elliptical, so that they move toward each other and then away from each other over periods of many millions of years.

If these two giant galaxies were alone in space, they would keep up the dance forever, or until the whole universe came to an end. However, the remaining galaxies in the Local Group all exert a gravitational pull, so that the Milky Way and Andromeda galaxies follow a somewhat complicated path that may occasionally make their approach a little too close. In short, they may collide and, in the long run, surely will collide. (There are examples of colliding galaxies among the many millions we can see in the sky.)

What will happen then?

The galaxies are not solid objects, to be sure, but merely clusters of many billions of stars. These stars are so far apart and are so small compared to the distances between them that, if the two galaxies strike each other a glancing blow, nothing much will happen. The stars of one will move among the stars of the other, with virtually no chance of collisions, and will affect each other only slightly. Eventually, the two galaxies will pull apart and go their own way.

But what if the two galaxies collide "head-on," so to speak, so that the center of one slowly approaches and merges with the center of the other?

The stars are much more densely packed in the center, allowing for far greater chances of stellar collisions. Worse yet, astronomers are now just about convinced that there is a black hole, as massive as millions of ordinary stars, at the center of each of the galaxies. The black holes will each swallow up many thousands, or even millions, of stars as they progress through each other's midregions and will finally merge, setting up a giant gravitational field that will continue to pull in stars.

This will involve an enormous quantity of radiation. The combined center of the two galaxies will emit radiation equivalent to that of a hundred or more galaxies of the ordinary type. In short, the two galaxies may become a *quasar,* the kind of superbright object that was more common in the younger days of the universe and that we can still detect billions of light-years away.

The radiation of the new quasar will heat up the thin gas that exists between the stars and drive it out of the galaxies. This means that no further new stars can be born, and the two galaxies will be forced to settle down to a steady aging process.

The quasar radiation at the center of the galaxies will be 30,000 light-years away from Earth, because we exist (very fortunately) in the outskirts of our galaxy. This means that the radiation will have thinned out considerably by the time it reaches us and will be stopped completely by our atmosphere. We will be able to see a very bright star in the sky in the constellation Sagittarius. That will be the quasar at the center, not masked by the clouds of dust and gas that now exist between us and it. Also, it may make space travel riskier.

Even if the collision should rip the galaxies apart and send our Sun careening outward into the intergalactic spaces, this won't affect us. Earth will just move out into those spaces along with the Sun and the other planets. The stars in our sky will gradually dim and disappear, but life will go on and we won't feel a thing.

Of course, if Andromeda's center should chance to head in our direction—! Happily, it is calculated that such a collision won't happen for some 4 billion years, so there is no immediate cause for alarm.

Ten Billion Light-Years Away

Finding something new in the sky is always exciting, but finding something new at an enormous distance—as astronomers have in the form of a "double quasar"—is doubly so. Anything we see at such a distance existed in the youth of the universe and astronomers simply hunger to find out as much as they can about those early days.

In 1963, the study of the far distance began when quasars were first discovered. These look like faint stars, but they were found to be at distances of 1 billion light-years or more. Hundreds of them have now been detected, and some are 10 billion light-years away.

We can't see much farther than that. This is not because we are penetrating to the end of the universe (there is no end) but because as we look at things farther and farther away, we are looking farther and farther back in time. A quasar that is 10 billion light-years away is seen as it was 10 billion years ago, when the universe was quite young. If we could penetrate farther we might be viewing a universe in which galaxies had not yet formed and in which clouds of hot radiation would be seen only as an opaque fog.

For quasars to be visible at such a distance, they cannot be stars as they seem, but must be whole galaxies. An ordinary galaxy, such as our own Milky Way, would not be visible at that distance, but quasars have enormously active centers that for some reason blaze with a light one hundred times as intense as that of ordinary galaxies. The result is they can be seen at even the farthest distances we can penetrate.

There seem to have been more such superactive galaxies long ago than there are now. Are young galaxies more likely to be quasars? What makes the centers so bright? What is the source of energy that powers them? What happens to a quasar when it finally "burns out"? Astronomers have many questions about quasars they would like answered.

For nearly twenty years, quasars were seen only as single objects; then in the early 1980s, an occasional "double quasar" was seen. Two quasars would be detected that were very close to each other. By then, the techniques for studying quasars by radio telescopes as well as by optical ones had been much improved, and their light could be analyzed in detail.

It turned out that the light from each of these two very closely neighboring quasars was identical in all characteristics. It was not as if there were two separate quasars, but a single quasar that, for some reason, was seen double. How could that be?

The logical answer was that the light from a quasar, as it traveled toward us, passed an ordinary galaxy that existed between it and us. This ordinary galaxy was too dim to see, but its gravitational pull bent the quasar's light a tiny bit. The light was bent inward on either side of the galaxy so part of the light crossed the other part very slightly and ended up in our telescopes as a closely spaced double beam. We therefore see two quasars where there actually is only one.

This effect is called a gravitational lens because it produces the same effect that an ordinary lens would. Albert Einstein predicted that such a phenomenon might exist seventy years before it was discovered.

Several examples of such gravitational lenses are now

known, and they can be used to make deductions about the galaxies that produce the effect even though those galaxies cannot actually be seen.

A certain quasar, listed in the catalogues as PKS 1145-071, has been known for years and is about 10 billion light-years away. It had always seemed just another quasar, but in December 1986 it was found to be double. There were two quasars there, very close together. It was naturally assumed to be another case of a gravitational lens.

In 1987, astronomers at the Multiple Mirror Telescope in Arizona analyzed the light from each of these closely spaced quasars. For the first time, the light turned out to be not quite identical in both. There were distinct differences that indicated that the quasars were not a single object seen double but two separate objects.

If this observation holds up, then PKS 1145-071 would be the first case that has been discovered of a true double quasar. It would represent two galaxies with enormously active centers at sufficiently close quarters to be circling each other.

If this is so, it is possible that these two galaxies do not exist all by themselves. There may be other galaxies in the vicinity that are not quasars, that do not have superactive centers, and that are therefore invisible. In short, there seems to be a distinct possibility that we are looking at a cluster of galaxies.

Galaxies do exist in clusters. Our own Milky Way is part of a cluster of two dozen galaxies, and other clusters with thousands of members are known. However, the new discovery would show that such clusters may already have existed 10 billion years ago. If so, this would force astronomers to sharpen some of their views on how galaxies formed in the first place.

Seeing the Past

We cannot see things as they are, no matter what we do. It takes time for light to travel from an object to our eyes, so that we always see things as they were in the past, never as they are right now.

Under ordinary conditions, this is not really important. If you see a friend across the street, you are seeing him as he was a hundred-millionth of a second ago, and that might as well be now. Once you leave Earth and begin to look at the heavenly bodies, however, the situation is more noticeable. It takes light one and one-quarter seconds to reach us from the Moon, so, as long as we remain on Earth, we always see the Moon as it was one and one-quarter seconds ago.

It takes light eight minutes to reach us from the Sun, so we always see the Sun as it was eight minutes ago. If some magical destruction suddenly wiped out the Sun, we would remain in blissful ignorance of that for a while and would continue to bask in sunlight and remain aware of the Sun, as if it were untouched. It would take fully eight minutes for the last of its light to reach us, and only then would we be plunged into darkness and know that the Sun had vanished.

The conditions are far more extreme for the stars. The distance light travels in a year (5.88 trillion miles) is a *light-year*, and the nearest star, Alpha Centauri, is 4.3 light-years away. This means that light takes 4.3 years to travel from Alpha Centauri to our eyes, and we always see that star as it was 4.3 years ago.

You might think that this can scarcely matter, because

Alpha Centauri was probably just about exactly the same 4.3 years ago as it is now. That's true, for stars change very slowly. Other heavenly objects are farther away. We see the star Sirius as it was 8.8 years ago, and the star Arcturus as it was 40 years ago.

In fact, we get radio signals from the very center of our galaxy (the huge conglomeration of 200 billion stars within which our Sun and we exist). So huge is our galaxy that those radio signals take 30,000 years to reach us. We can only be aware of the center of our galaxy, then, as it was 30,000 years ago.

And, of course, there are other galaxies far outside our own. The nearest large galaxy is the Andromeda galaxy (the farthest object we can see with our unaided eye). It is 2.3 million light-years away. This means that when we study the small bit of haze we see when we look at the Andromeda galaxy, we are looking at it as it was 2.3 million years ago. Nothing will allow us to see it more recently than that.

There are galaxies we can observe much farther than the Andromeda: tens of millions and even billions of light-years away. When we get to such distances, we are looking so far into the past that there has indeed been time for enormous changes to have taken place even in objects as long-lived as stars and galaxies. We are managing to see them as they were when they were comparatively young. Unfortunately, the farther the object, and the farther into the past we probe as we look at it, the dimmer it is and the less detail we can see. (You can't win!)

The universe had its beginnings, scientists estimate, about 15 billion years ago, and the billions of galaxies that now exist must have gotten their start in the first few billions of years after that beginning. If, then, we want to see things as they were at the beginning, we must look at objects that are billions of light-years away. Even the largest galaxy is represented by only a tiny bit of radiation at such a distance.

Early in January 1987, Hyron Spinrad, an astronomer at the University of California in Berkeley, announced that just such an object had been observed. It was a galaxy known as 3C 326.1, and it was no less than 12 billion light-years away. This means it can be seen as it was 12 billion years ago, when it was young, and presumably just forming. It is the first time astronomers have ever noted a large galaxy in the process of being born.

To be sure, all they get is the tiniest spark of radiation in the most advanced radio and optical telescopes now in existence. Yet by closely analyzing the radiation they get, they can tell that the young galaxy consists of a huge, hot cloud of gas about three times as wide as our own galaxy. It would also seem that at least a billion stars have already formed within it. Presumably hundreds of billions more will eventually form (or, in fact, have actually formed billions of years ago—except that their light hasn't reached us yet).

There are also radio waves reaching us from this young galaxy, and these may be sent out from a black hole at the galaxy's center. Astronomers speculate that many black holes formed at the very beginning of the universe, when the big bang took place, and that these serve as nuclei around which galaxies form. This young galaxy and other galactic babies we may observe in the future may help put flesh on such theories.

The Fastest Telescope

Columbia University is planning to build a completely new kind of telescope, one that won't specialize in being big but in being fast.

It will have a mirror that will consist of one thousand pairs of surfaces, each about an inch or so across, and each housed inside tubes the size of saltshakers that are held in place by solid-state magnets. All the surfaces are kept in exact coordination by a robot device, and each one is designed to spread out the light of a particular star into a rainbow (or *spectrum*) that can be studied in detail. The telescope has an estimated cost of $30 million.

An ordinary large telescope can concentrate on a patch of sky only twice as big as the area taken up by the Moon. The Columbia fast telescope, on the other hand, will be able to study a patch one hundred times as large. An ordinary large telescope can usually study the spectrum of one star, or other astronomical object, at a time. The Columbia fast telescope, on the other hand, can analyze the spectra of one thousand different objects at a time.

This is important, for the spectrum gives us an enormous amount of information about an astronomical object. It tells us its chemical composition, its surface temperature, the speed at which it is moving toward us or away from us, its magnetic properties, and so on.

The spectra are particularly important in connection with the galaxies, which are scattered over billions of light-years all through the visible universe. Each galaxy is made up of many billions (sometimes trillions) of stars.

All the distant galaxies are receding from us because the universe as a whole is expanding. The faster a galaxy is receding, the more distant it is. Because the spectrum of a galaxy tells us how fast it is receding, it therefore tells us how distant it is.

If we had the spectra of all the galaxies and knew the distance of each, we could build a three-dimensional model of the universe and see how the galaxies are distributed. This might help tell us how the galaxies were formed, and this would in turn tell us much about the youth of the universe, which in turn might give us information about its beginning and its possible end.

There are perhaps 100 billion galaxies in the universe altogether, although the vast majority are so far away and so dim their spectra cannot be studied. But there are at least 2 million galaxies close enough to us to be studied in detail. Spectra of these nearby galaxies have been taken and studied for three-quarters of a century, but in all that time only 7,500 galaxies have been adequately studied and have had their distances determined.

Even this is enough to give astronomers a hint that the galaxies are arranged in a complex and puzzling manner, but we must know many more distances if we are to have a chance of grasping and understanding the arrangement. Astronomers are hoping to double the number of galactic distances known, a project that would take nine years with ordinary telescopes. The Columbia telescope, however, once it was built, would be taking one thousand spectra at a time and would double the number of known distances in *one week*. In two years, it might determine the distance of a million galaxies and multiply the volume of examined space five hundred times. How much more we would then know about the universe!

Another great puzzle about the universe is the *missing mass*. There are indications that all the mass we can detect in the universe is only 1 percent or less of the total mass. The amount of mass that is present in the universe dictates what its course of history and its final ending will be, but we cannot be sure of this course and this end until we know what the missing mass is.

Our own Milky Way contains about 200 billion stars, but it too may have its share of missing mass. It would help if we knew the precise distribution of all the stars in our galaxy, but again this requires knowing distances, speeds of motion, and other details of many millions of stars. Ordinary telescopes simply can't undertake such a task without spending many years on the job.

The Columbia fast telescope could quickly supply the necessary data that would make it possible to understand the true organization of our galaxy and perhaps give us an insight into what the missing mass might consist of and where it might be located.

In addition, a truly copious study of the spectra of numerous stars would give us detailed information of the chemical composition of each. The chemistry of the universe is constantly changing because, at the core of stars, heavy elements are being built up. Supernovas spew forth these elements into the cosmic dust and gas clouds, and from this material new stars form.

If we knew enough about the present chemistry of the stars we might be able to deduce just how the elements were built up and get an idea of the course of development of the galaxy. In a few years of operation, the Columbia fast telescope might give us details of galactic structure almost beyond present imagining, and we might know much more about the origin of the Sun, the Earth—and ourselves.

The Oldest Birthday

The oldest object in the universe is, obviously, the universe itself. The universe has the oldest birthday, but how old that birthday is is still a matter of question, and yet another new estimate was made in 1987.

It was not until 60 years ago that anyone had any idea at all how old the universe might be. Even if the Earth itself is 4.6 billion years old (a figure scientists now feel quite certain about), the universe might have existed for uncounted aeons before its formation.

But then, in the 1920s, Edwin P. Hubble estimated the distances of various galaxies from Earth and the speed at which they were receding from us. (They were very nearly all receding.) It seemed quite clear that the farther away the galaxy, the faster it was receding—and at a rate proportional to its distance. This universal recession in just this fashion was most easily explained by the theory that the universe, as a whole, is expanding.

If one imagined time's reversing itself (as in a backward-running film), then all the galaxies would seem to be approaching each other until, at some point, they would smash together into one vast conglomeration.

It was for this reason that, in 1927, a Belgian astronomer, Georges Lemaître, suggested that the universe began as a tight conglomeration of matter that exploded. The galaxies are now still receding from each other as a result of that initial explosion, which is now popularly called "the big bang."

In 1929, Hubble suggested that the big bang took place 2 billion years ago. This created quite a stir because geologists were quite certain that the Earth was considerably older than this (more than twice as old, we now know). Hubble's estimate would mean the Earth is older than the entire universe.

Astronomers felt there was no arguing with the galaxies, so there was an impasse until 1942. At that time the German-American astronomer Walter Baade took advantage of the wartime blackout to study the Andromeda galaxy (the nearest large galaxy) in detail. It turned out that one of the devices used by astronomers to judge the distances of the galaxies had certain unexpected complications. Taking those complications into account, all the galaxies turned out to be up to three times as far from each other as had been thought. That meant that, if the film of time were reversed, it would take at least three times as long for the galaxies to come together, so the big bang occurred at least three times as long ago as had previously been calculated. With a new minimum age of the universe of 6 billion years, the geologists were satisfied.

It didn't stop there, however. Astronomers kept making more refined measurements of the rate at which galaxies were receding and making subtle observations that told them how old individual stars might be. The result is that, nowadays, astronomers are satisfied that the universe is somewhere between 10 and 20 billion years old. The figure most often quoted (by me, for instance) is halfway between these two figures, or 15 billion years old. Most astronomers feel, indeed, that the true figure is likely to be more than 15 billion years rather than less.

The astronomer Harvey Butcher, at the University of Groningen in the Netherlands, has tackled the problem from another angle. Analyzing the light of a particular star can tell us what chemical elements are present in it. Some of these elements are radioactive; the ages of these stars can be estimated from the amount of elements present that are produced by such radioactivity. Some stars seem to be about as old as the universe and must have been formed soon after the big bang.

Butcher has studied the elements thorium and neodymium in these very old stars. Under the conditions within the stars, thorium should break down to neodymium at a certain rate. Butcher calculated that the breakdown had been proceeding for just about 10 billion years. This would mean that the stars themselves were not more than 10 billion years old, a figure he announced in July 1987. Since the stars are among the oldest we can see, the universe may not be more than 11 or 12 billion years old.

However, Butcher's measurements are extremely delicate, and it may be that the stars he studied are not actually the oldest. Astronomers therefore reacted to the new estimate with caution. The age of the universe, to the nearest billion years, is still up for grabs.

Superstars?

It often happens that an important observation or theory by a great scientist of decades or centuries past has to be extended or modified. But every once in a while, it turns out that the extension must be abandoned and we find that the scientist was originally right after all. Such a case of back-to-where-we-were took place in 1988.

It began with the British astronomer Arthur S. Eddington. In the 1920s, he asked the question, Why doesn't the enormous gravitational pull of a star like the Sun simply force it to collapse into a tiny ball of crushed atoms?

The answer seemed to be that the internal heat of the Sun kept it expanded against the pull of gravity. Eddington set about working out the balance between gravitational pull and internal heat and deduced that the Sun's core had to be at a temperature of millions of degrees C. This was important in explaining the nature of nuclear reactions within the Sun and the way it and other stars obtained the energy necessary to keep them shining for billions of years.

Eddington found that the more massive a star, the more intense its gravitational pull inward, and the higher the core temperature had to be to balance that pull. By the time the star was somewhere between sixty and one hundred times the mass of the Sun, it would no longer be possible to maintain a balance. To prevent the star from collapsing, the internal temperature would have to be so high that the star would actually explode. Therefore, Eddington concluded, stars with a mass considerably more than sixty times that of the Sun could not exist. And for over half a century, indeed, there was no reason to think he was wrong. Stars with greater mass were not found.

But then, in the 1980s, stars that seemed to have a mass several hundred times that of the Sun, even more than one thousand times the mass of the Sun, were discovered. How were such "superstars" possible? Eddington's work had to be gone over and modified to account for these huge stars. (Several years ago, in fact, I wrote an essay about these superstars and how they were changing our views of stellar physics.)

But then the superstars fell apart—almost literally.

For instance, there is a star in the Large Magellanic Cloud called Sanduleak. Its distance was known to be about 160,000 light-years, and it was so bright at that distance that it had to have a mass at least 120 times that of the Sun to produce all this light.

It was, however, observed and photographed with newer and better telescopes early in 1988. The image of the star was then analyzed with updated techniques to see how the brightness varied from point to point. It turned out that the star was not evenly bright and was therefore not a single star. It was, actually, a very closely spaced cluster of at least six stars. At the great distance of Sanduleak, this cluster seemed to melt together into a single star as seen under ordinary telescopic conditions.

Other such very bright and therefore very massive stars have also, by this technique, been resolved into tight groups of stars—and no one star in any of these groups seems to have a mass more than sixty times that of the Sun. In other words, Eddington was right all along, and the superstars have vanished from the sky.

Does this have any importance aside from the fact that it allows Eddington to rest peacefully in his grave? As a matter of fact, it does. For one thing, it once again demonstrates that scientists must constantly be probing and testing their conclusions and that their findings can be subject to change.

In this case, the confirmation of Eddington's theories had relevance beyond the existence or nonexistence of superstars. The discovery of star clusters, in turn, caused scientists to reevaluate their estimates of the distances of galaxies from Earth.

It is important for astronomers to estimate the distance of faint galaxies in order to get a general notion of the overall size of the universe. To do so they try different techniques, working out the distance of the nearer galaxies, using that distance to work out the distance of farther ones, and so on.

One technique has been to study those galaxies that were so near that individual stars could be made out within them but so far away that the only stars that could be seen were the ones that were the very brightest in the galaxy. It was assumed that these "very brightest" stars in these distant galaxies delivered as much light as the very brightest star in our own galaxy. We knew just how far away and how bright very bright stars were in our own galaxy, so it was possible to calculate how far away distant galaxies were by working out how far they would have to be for their brightest stars to seem no brighter than they were.

But it may be that we have been deluding ourselves. It may be that whereas the brightest stars in our galaxy are seen clearly enough for us to be sure they are single stars, the brightest stars in distant galaxies are actually clusters that, taken together, shine much more brightly than single stars can.

If this is so, then some distant galaxies may be two or three times farther away than we had thought, far enough that a cluster of stars shines at about the same brightness as a single one would if it were closer. In that case, the universe is much larger than we thought and much older, and this would send astronomers back to their drawing boards.

The Baby Pulsar

On January 18, 1989, astronomers finally detected something for which they had been searching for two years.

Two years earlier, a star in the Large Magellanic Cloud was seen exploding, becoming Supernova 1987A. According to theory, some of it should have collapsed into a neutron star. This could be a tiny body about sixteen miles across that would still have the mass of something like our Sun. It would be spinning very rapidly and it could be detected because it would send out beams of light and other radiation in pulses, one for each time it rotated. For that reason a neutron star is also called a "pulsating star" or, for short, a *pulsar.*

Pulsars were first discovered in 1969, and the first one studied rotated in one and one-third seconds, or about three-quarters of a turn per second. This was astonishingly fast. The Earth rotates in twenty-four hours, and considering that it is eight thousand miles across, a point on its equator moves at the rate of about 1,035 miles per hour, or a little more than a quarter-mile per second.

Jupiter, which is 88,000 miles across, rotates in 9.9 hours, so that a spot on its equator moves at a speed of about 7.4 miles per second. If Earth rotated at this rate, it would fling equatorial matter into space, but Jupiter has a more intense gravitational pull.

The first pulsar discovered, however, rotated so rapidly that, even allowing for its tiny size, a point on its equator was moving at the rate of a little more than forty miles per second.

Astronomers quickly decided that a pulsar's spin must gradually slow over time. Therefore, a young pulsar must spin more rapidly than an older one. The youngest pulsar we knew (before 1989) was the one in the Crab Nebula that is only nine hundred years old. And, sure enough, it spun at a rate of thirty times a second, or about forty times as fast as the old pulsar that was the first to be discovered. A point on the equator of the Crab Nebula pulsar would move at about sixteen hundred miles per second. Only the enormous gravitational pull of a pulsar (perhaps twenty-five billion times that of the Earth) could hold an object together at that rate of spin.

But then, in 1982, astronomers discovered a pulsar that spun at a rate of 642 times per second, and it was an *old* pulsar. It made one turn in a little more than a thousandth of a second (or millisecond). A point on its equator must be moving at a speed of 32,000 miles per second, or at a quarter the speed of light. Other such millisecond pulsars were discovered, and they were usually found near another star. Such pulsars picked up material from the nearby star, which caused them to speed up, and sometimes they absorbed the companion star altogether.

But how fast does a newly formed pulsar, a baby pulsar, spin? As soon as Supernova 1987A appeared, astronomers began to hope to be able to see such a baby pulsar. Unfortunately, the debris in the outer regions of the huge explosion hid the center where the pulsar would be. Only now has the fog lifted slightly and made it possible to detect the pulses, and it turns out they are generated at a rate of 1,969 times per second. The baby pulsar is spinning in half a millisecond. This is twice the rate that even the most daring astronomer predicted.

A pulsar that is turning at 1,969 times per second has a

point on its equator moving at a speed of 100,000 miles per second, or better than half the speed of light. This is astonishing, for even the enormously intense gravitational field of a pulsar can hardly hold it together when it is spinning at this speed.

And this isn't even the most astonishing part of the discovery. It was also found that the brightness of the pulsar fluctuates somewhat in a period of eight hours. This most likely means that it has a companion object with perhaps one-thousandth its own mass, that is, with the mass of Jupiter. The two are rotating about each other, with one revolution every eight hours.

The pulsar and its planet, however, are so close together that one wonders how the planet could have survived the explosion. In fact, the planet is so close to the pulsar that before the explosion, it must have been inside the outermost layers of the star.

One possible explanation was that two years ago, when the pulsar first formed and was a newborn infant, it was spinning even faster than 1,969 times per second and didn't hold together. A little piece might have spun off it and carried away some of the energy with which the pulsar spins. What was left behind spun more slowly and was able to hold together.

The rapid spin also raises questions about how a pulsar can be so bright, how strong its magnetic field must be, and so on. However, astronomers only caught the one glimpse for a short period of time, and then the clouds about the pulsar thickened again. The astronomers are still waiting for additional looks at times when they will be able to detect the pulsar's properties more clearly. Some of the puzzles may clear up with a better look—or become still more puzzling.

Beyond the Beyond

Last autumn Voyager 2 passed Neptune after a twelve-year journey (so far) and is heading out far beyond. It is carrying a recording that tells about Earth and includes sights and sounds of our planet. This has frightened some people who feel that we are giving away our location to alien creatures from other worlds who might come to conquer us.

Those who think so don't understand the size of the universe or the chances that Voyager 2 will be found by anyone.

It took twelve years for Voyager to go from Earth to Neptune, and now it is passing beyond. Where will it go hereafter? What worlds will it reach? Voyager 2 is drifting under the diminishing strength of the Sun's gravitational field (as it recedes from the Sun) and the vanishingly small effects of the gravitational fields of different stars. We can use these gravitational effects to make allowances, so we know exactly where Voyager 2 will go.

We know all the stars in our neighborhood, and Voyager 2 probably is not going to hit any of them. Of course, there may be dark bodies we don't know about, a wandering planet or an asteroid that Voyager 2 might smash into, but the chances of that are so tiny, there's no use even thinking about it.

The Sun emits a "solar wind," a spray of charged particles in every direction. The spray becomes thinner and thinner as it recedes from the Sun until it fades away into interstellar space. Voyager 2 will pass beyond the reach of the solar wind in the year 2012.

In the year 8571 (nearly 6,600 years from now), Voyager 2 will be 0.42 light-years from the Sun. That's about 2.5 trillion miles. However, even the nearest star is ten times farther away. At that point, Voyager 2 will make its closest approach to Barnard's Star, which is 5.9 light-years away from us (35 trillion miles) right now. Voyager 2 will be only 4.03 light-years from it (24 trillion miles). Having skimmed by, if you want to call that skimming, it will pass on.

In the year 20,319, Voyager 2 will be 1 light-year from the Sun (5.9 trillion miles), and it will make its closest approach to Proxima Centauri, which is the nearest star to us. Proxima Centauri is 4.3 light-years away from us (25 trillion miles) but, of course, Voyager 2 is not heading in its direction. It is moving well to one side and its closest approach is 3.21 light-years (19 trillion miles).

A mere 310 years later, Voyager 2 will make its closest approach to Alpha Centauri, a double star just a little farther off than Proxima Centauri. That closest approach will be 3.47 light-years away (20 trillion miles.)

All this time, you must understand, Voyager 2 is sufficiently close to the Sun so that it keeps very slowly spiraling about it in response to its gravitational pull. It is still inside the solar system. Far beyond the farthest planet we know of, Pluto, there may be one or two other planets, but so far there have been no signs of them. We are pretty sure, however, that way out there are a hundred billion or more small icy bodies—comets. This is called the Oort cloud after the astronomer who first theorized they were there.

Voyager 2 will enter the Oort cloud about the year 26,262 and it will continue to move through that cloud for about 2,400 years. It might seem to you that if Voyager 2 moves through a region that contains a hundred billion icy bodies, each at least a dozen miles across, it is bound to strike one and be destroyed.

Not so. The volume of the Oort cloud is so huge that even with a hundred billion bodies slowly circling within it, the chances that Voyager 2 will strike one are virtually zero. About the year 28,635, Voyager 2 will leave the Oort cloud and be in interstellar space.

After a million years of traveling, Voyager 2 will be about 50 light-years from the Sun (which as star distances go is still pretty much in our own backyard). In all that time, the closest it will have approached to any star at all would be its pass at Proxima Centauri, where it was only 3.21 light-years away. In a million years, it will never come closer than 19 trillion miles to any star and the chance that any alien creature is going to come across this little, silent probe far off in the depths of space between the stars is absolutely too small for us to worry about.

But then, in that case, why did we send a message if there is virtually no chance that it will ever be picked up?

Remember, a million years is a short time in the history of the universe. The universe has already endured 15,000 times a million years and it will surely continue to exist. Someday, undoubtedly long after we're gone (for the chance that humanity will endure for even a single million years is, frankly, not great), someone may come across it.

But who cares, if it's long after we're gone? Well, think about it. Do we want to disappear without a trace? Don't we have a little pride in the human species? Surely, we would want other intelligences to know we were once here and what we managed to do.

Why Are Things as They Are?

In November 1988 a high-powered scientific meeting was held on a topic that scientists have been discussing for years: the anthropic principle.

Anthropic is from Greek and means "concerned with man." The anthropic principle tries to maintain that human beings, as observers, are necessary to the very existence of the universe.

It might seem that the opposite is true. Here we are on a small planet of an average star lost in a galaxy that contains hundreds of billions of stars, with additional stars in a hundred billion other galaxies. Why should there be so unimaginably huge a universe just for us?

The answer is that the smaller the universe, the less time it takes for it to expand and then contract out of existence. The universe must be as huge as it is in order for us to have had time to evolve.

In addition, the laws of nature are such that atoms can form. If these laws were slightly different, the formation of atoms would be impossible. Again, the events after the big bang seem to have been such as to allow stars and galaxies to form. Slight differences would have made them impossible. If it weren't for atoms' and stars' and galaxies' just happening to be possible, we ourselves would not be possible.

Even on Earth, a slight change in Earth's orbit or in the Sun's mass, and Earth would not be habitable. Even if it were, small changes in chemistry—for example, if water did not expand when it turned into ice, or if carbon atoms didn't manage to hook on to each other—would have made life impossible.

Quantum theory also makes it look as if we are indispensable. According to quantum theory, there are conditions in which it is impossible to tell just what an electron is doing until it is actually observed. When the electron is not observed, it is not even theoretically possible to decide what it is doing. Some scientists take this to mean that the universe can't exist without observers.

A universe must have observers, according to this theory, and it must have observers from the start to the end. But then, even the simplest human beings didn't evolve till the universe was 15 billion years old. Did dinosaurs qualify as observers? The Earth itself wasn't formed till the universe was 10 billion years old. Does that mean there are other forms of life on other planets that did the observing? Or does it mean that the universe was formed just for the benefit of human beings by God? And that God is the universal observer through all of eternity? This postulation might seem necessary according to the "strong anthropic principle."

However, most scientists prefer a "weak anthropic principle." To see what this means, consider this question: Why do your ears have the shape and position they have? The answer might be so that spectacles will fit over them. In that case, ears must exist and must be where they are, and it is the existence of spectacles that determines that.

But it's the other way around. Spectacles were designed to fit the ears, not vice versa. If ears were located elsewhere or didn't exist at all, spectacles would have been designed in a different way.

In the same way, there may be an indefinitely large number of universes in existence, each with a different set of laws of nature. In perhaps all but one of these umptillions of universes, the laws of nature don't allow life to exist. In only one of them do the laws of nature allow for the existence of life.

This one universe would be ours, and we would have evolved in it and then marveled at how exactly suitable the universe is for us. But this has nothing to do with us, really. We find our universe perfect only because it is the only one we could exist in. Maybe, in other universes where life (as we know it) could not exist, other kinds of life or other types of unimaginable phenomena might prevail. And every one of these lives or phenomena that had the capacity to wonder would wonder why their universes are so fit for them.

How can we decide whether this weak anthropic principle is correct? After all, our own universe is the only one we can observe. An Italian scientist, E. W. Sciama, has made a suggestion.

If there are an indefinite number of universes, there may be a great many that are close enough to perfect to allow our kind of life to exist. Ours would be just one of them, and it might not be the most nearly perfect.

If we knew more about our universe, if we could make measurements that are more delicate than we have so far made, if we could learn more about life and its requirements than we at present know, then perhaps we could see that our universe is not completely perfect. We might even manage to design (in the mind) a universe that would be more suitable than ours by modifying the precise form of this natural law or the precise value of that constant.

If our own universe were a little imperfect, then it would be more likely that there is a tiny range of universes that would be suitable for us. This would make the weak anthropic principle seem a little more likely and would be a point against the strong one.

Where the Universe Ends

How far is far? Astronomers may have already seen objects that are 17 billion light-years away: about 100 billion trillion miles away.

This is not bad really. As recently as 1920, astronomers thought our own Milky Way galaxy and some smaller neighboring objects were all there was to the universe; the most distant objects were only 150,000 light-years away.

But then, in the 1920s, it became clear that there were other galaxies, many other galaxies, billions of them. What's more, the universe was expanding, so that groups of galaxies were moving steadily away from each other. The waves in the light from a galaxy moving away from us are stretched, making the light seem redder in color. This is called a *red shift* and can be measured by the position of certain dark lines in the wave pattern (spectrum) of the light. The greater the red shift the farther away the galaxy.

By the 1940s, it was quite clear that even the nearest large galaxy outside our own was more than 2 million light-years away. More distant ones were hundreds of millions of light-years away. Beyond that there might still be many other galaxies, but at greater distances they became too dim to see.

In the 1950s, it turned out that certain objects that looked just like ordinary stars emitted radio waves in unusual quantities. When they studied these objects, scientists were unable to identify the dark lines in their spectra. In 1963, the scientists realized that the dark lines were red-shifted out of position by an enormous amount, meaning the objects must be very far away.

These strange stars were called quasars. These quasars turned out to be very distant galaxies with centers that blaze with light for some reason. They are so far away that we can't make out anything but these blazing centers, which make them look like stars.

Even the nearest quasar is 1 billion light-years away. Other quasars are much farther, as much as 10 billion light-years away or more. We now know that large numbers of quasars exist in every direction, but it is not easy to detect them against the even larger numbers of ordinary stars that fill space.

When we look at an object that is 10 billion light-years away, we are looking at light that has taken 10 billion years to reach us. We therefore see the object as it was 10 billion years ago, when the universe was only half its present age, perhaps. Apparently, quasars formed in vast numbers in the early days of the universe, reaching a peak about 13 billion years ago and then declining in numbers ever since, as fewer and fewer new ones formed and more and more old ones faded out. So the study of very distant (and therefore very old) quasars should give us useful information about the youthful days of the universe.

One way of expressing the distance and age of a quasar is by measuring how much its light wavelengths have been stretched. If the wavelengths are double what they should be, this is a red shift of 2; if triple what they should be, this is a red shift of 3; and so on. The higher the number, the farther and older the quasar.

Until recently, the greatest red shift observed was 3.8, corresponding to a distance of about 15 billion light-years. Astronomers suspected that they wouldn't be able to detect farther ones because the universe might not have formed galaxies earlier than that.

They were wrong. In September 1986, a quasar that had a red shift of 4.01 was detected. In 1987, a number of different quasars with red shifts of more than 4 were detected. The current record holder has a red shift of 4.43. These are perhaps 16 billion years old.

The light waves from the most distant quasars are stretched so much that more and more of that light is in the infrared. The light we can actually see is very dim. Generally we detect such quasars only because their radiation is rich in radio waves.

So what if scientists searching for even more distant quasars focused on finding objects whose spectra are rich in infrared light and also had very high red shifts? At the University of Arizona, a team under the leadership of Richard Elston has used arrays of powerful infrared detectors for just this purpose.

In January 1988, they reported that they had located objects rich in infrared light that seemed to have unusually high red shifts: some as high as 6, perhaps. It seemed from what information they gathered that these objects were galaxies in the process of formation and that they were at least 17 billion light-years away.

At that time the universe may have been only 2 or 3 billion years old. If this is when galaxies were forming, we can't expect to see anything farther away, because there would be nothing to see—just a haze of energetic matter that had not yet collapsed into galaxies. We will have come to the end of the universe because we will have come to the beginning of the universe. The two are the same.

INDEX

Absolute zero, 48, 50, 83, 94
Acid rain, 29, 175, 177, 180
Adenosine triphosphate (ATP), 103–104
Aepyornis ("elephant bird"), 162
Africa, 11, 144, 167, 201
African elephant, 161
Aging, 72–74
Albatrosses, 158, 160–61
Algae, 181, 183
Alloys, 50
Alpha Centauri, 355–56, 371
Alpha-aminoisobutyric acid, 149
Amber, 163, 165
Amino acids, 96–97, 105, 106, 127; found in sediments, 148–51
Amphibians, 31–32, 137–38, 143, 144
Anders, Edward, 179, 238
Anderson, Carl, 81
Andrews, Thomas, 58
Andromeda Galaxy, 306, 341, 349–50, 351, 356
Angular momentum, 271–72
Animals, 144, 160–62
Antarctica, 184–86, 200; dinosaurs in, 142–43, 144–45; meteorites found in, 234, 285–87; ozone hole, 181–83
Antarctic Ocean, 184
Anthropic principle, 373–75
Antibiotics, 104
Antielectrons, 81
Antigalaxies, 314–15
Antileptons, 78, 79
Antimatter, 52, 78, 81–83, 215, 313–16
Antineutrons, 315, 316
Antiparticles, 81–82
Antiprotons, 52, 82–83
Antiquarks, 79
Antistars, 314–15, 316
Apollo 16, 264
Apollo 17, 264
Archaeobacteria, 135
Archaeopteryx, 151–53, 154–56
Arctic Ocean, 200, 203–204
Arctic region, 200
Argon, 8–9, 45, 46
Armstrong, Neil, 205
Artificial intelligence, 113
Ascension Island, 166–69
Asia, 144; India's collision with, 121, 144, 204–205, 212
Asimov, Isaac, 21, 237, 330
Asimov, Janet, 15
Asteroid belt, 289
Asteroids, 211, 235–37, 274–75, 288, 289; composition of, 289; in impact theory, 178–80
Astronauts, 259, 264, 266, 314
Astronomical unit (AU), 258, 312–13
Astronomy, 2, 305–306
Atlantic Ocean, 167, 212
Atmosphere, 253, 254; Earth, 125, 187, 188, 211; Mars, 301;

Atmosphere (*cont'd.*)
Pluto, 278; Titan, 229; Triton, 261; Venus, 226, 227–28
Atom smasher(s), 51–53, 63, 77, 326
Atomic clocks, 192–93
Atomic nuclei, 75–76, 77, 87–88, 334–35; particles in, 317, 318
Atomic numbers, 44
Atoms, 66–67, 69, 72, 224; combinations of, 45–47; formation of, 373; structure of, 71, 75–76; X rays of, 39–40, 41
Australia, 144, 169, 170
Australopithecines, 13–15, 16–18, 20
Australopithecus afarensis, 9
Australopithecus robustus, 16–17, 18, 27
Avery, Oswald Theodore, 187
Avise, John C., 167

Baade, Walter, 362
Bacteria, 99, 100–101, 126, 127, 134–35, 140–41; bioluminescent, 102–104; nitrogen-fixing, 176
Bacterial cells, 141
Bada, J. L., 127, 148, 149, 151
Baluchitherium, 161
Barnard's Star, 309, 371
Bats, 158
Bednorz, J. G., 63–64
Bees, 163, 165
Belton, Michael J. S., 290
Benzene molecule, 69–71
Beryllium, 47
Big bang, 53, 306, 318–19, 357, 361–62, 363, 373
Big bang theory, 319
"Big Game Hunting in Space" (Asimov), 237
Biological clock, 114–16
Bioluminescence, 102–104
Birds, 19, 32, 151, 155, 158, 161–62; primitive, 152, 154–56
Birthrate (human), 36, 74
Black dwarf(s), 279, 347
Black holes, 260, 330, 332, 334–36; detection of, 337–39; at galaxy centers, 350, 357
Blue whale, 162, 185
Boltwood, Bertram B., 8
Bosons, 77, 78, 79, 80
Bourgeois, Joanne, 179
Bowen, Brian W., 167
Brachiosaurus, 161
Bradley, James, 123
Brahe, Tycho, 305
Brain, C. K., 27
Brain size, 14, 16, 17, 18, 20
Brecher, Kenneth, 328
Broglie, Louis de, 108
Brown dwarf(s), 280–81, 347–48
Burrowers, 136, 138–39
Butcher, Harvey, 362–63

Calcium carbonate, 227, 286
Callisto, 253
Campbell, Bruce, 309
Carbon, 84, 240, 255, 287
Carbon atoms, 28, 29, 72, 239; in benzene, 69–71; central to life, 223–25; in diamonds, 84, 85, 86
Carbon compounds, 126, 149, 180, 225; in meteorites, 286, 287
Carbon dioxide, 29, 58, 125–26, 180, 223, 265; in atmosphere, 187–88, 189–90; *see also* Supercritical carbon dioxide
Carbon 14, 265–66
Carbon-14 dating, 195, 265, 266
Carbon monoxide, 180, 227
Carbonaceous chondrites, 149, 223–34
Carr, Archie, 167
Cell nuclei, 134, 140, 141
Cells, 142; first, 133–36; of green plants, 140–41
Cepheid variables, 340–41
Ceres, 275
Challenger, 241, 292
Chamberlain, Owen, 82
Chandler, Seth C., 123
Chandler wobble, 123–24
Charon, 273, 276–78

Chemical bond(s), 66–68
Chemical elements, 62–63
Chemical reactions, 96, 98, 105, 106, 186
Chemistry: of stars, 360, 362–63
Chimpanzees, 9, 13, 19
Chiron, 289–90
Chiu, Hong Yee, 329–31
Chlorine atoms, 182
Chlorofluorocarbons (CFCs), 181–82
Chlorophyll, 141
Chloroplasts, 140, 141–42
Christy, James W., 276
Chromosomes, 97, 105, 106
Circadian rhythms, 115
Climate, 29, 199–202
Coal, 28, 29, 30, 84, 189, 190
Coelacanth, 31–33
Coleman, Patrick, 167
Columbia University, 163, 202, 203, 358–60
Columbus, Christopher, 194, 197–99
Comet belt, 256–57
Cometary collisions, 219
Comets, 211, 215, 216, 217–18, 220–21, 225, 287, 289; angular momentum, 272; color of, 218–19; composition of, 215, 218, 220, 268, 289; iridium in, 178; large, 289, 290; Oort cloud, 270–71, 272; Sun-grazers, 267–69; tails, 220–21
Companion stars, 308, 310
Complex molecules, 59, 223, 225, 320; X rays of, 40–41
Computer chips, 113
Computers, 63, 221, 305
Continental climate, 200
Continents: movement of, 143–45, 203–204
Copper, 50, 64, 232
Coracoid bone, 156
Coronagraph, 269
Cosmic rays, 72, 91, 245, 246, 265, 315; from supernovas, 307
Crab Nebula, 368
Cram, Donald J., 96
Craters: Moon, 207–208; Venus, 213, 228
Crick, Francis, 106, 110
Critical mass, 87
Critical pressure, 57–58
Critical temperature, 57–58
Cro-Magnon man, 22
Crookes, William, 63
Crowley, Thomas, 201
Crust: Earth, 120, 121, 145, 180, 206, 211–13, 228, 233; Venus, 228
Cyanobacteria, 134–35, 141, 142
Cyclotron, 52
Cygnus X-1, 335–36

Damon, Paul E., 266
Dark nebulae, 320–21
Dating techniques, 8–9, 12, 195
Davisson, Clinton J., 108
"Day Is Done, The" (del Rey), 21
De Nova Stella (Brahe), 305
Death rate (human), 34, 36
del Rey, Lester, 21
Deoxyribonucleic acid (DNA), 73, 97, 108–10, 187; structure of, 40, 105, 110
Deuterium, 87, 89, 92
Diamondizing, 85–86
Diamonds, 84–86, 146, 238–40
Dinosaur fossils, 144
Dinosaurs, 142–45, 161, 374; extinction of, 121, 144, 145–48, 148–51, 178, 180, 219, 236
Dirac, Paul, 81–83
Dolphins, 19, 26, 31, 185
Donn, William L., 203–204
Double-planet(s), 277
Double quasar, 352–54
Down-quarks, 78–79
Dragonflies, 162, 165
Duckbill platypus, 170–71
Dust cloud (theory), 178–79, 180
Dust clouds, 223–24, 238–39, 308, 314, 319–22, 341, 360
Dutch elm disease, 100–101
Dwarf star, 257

Earth, 35, 219, 220, 233, 258; age of, 7, 8, 136, 233, 361, 362; asteroid close calls, 235–37; axis, 122; chemical structure of early, 125–26; climate, 199–202; in collision of galaxies, 351; development of life on, 223–25; end of, 298, 299; formation of, 223; habitability of, 373–74; internal heat of, 131–33, 211–12, 228; land life, 136–38; mass distribution of, 121, 123–24; Moon in history of, 205–208; orbit of, 203, 300–301; rotation of, 191–93, 226, 367; shifting, 119–21; wobbling, 122–24; *see also* Crust, Earth
Earthquake waves, 132
Earthquakes, 123–24, 131, 143, 191, 212; U.S., 128–30
Echidna, 170, 171
Eclipses, 192, 247–49, 276–78, 283
Ecological chain, 181, 183
Ecology, planetary, 185
Ecosphere, 300–301, 302
Eddington, Arthur S., 364–65
Einstein, Albert, 43–44, 259, 260, 322, 325–26, 328, 343, 353
Electric charges, 77, 78, 81
Electric sensors, 171
Electromagnetic interaction, 79
Electron beam(s), 40, 112–13
Electron microscope, 42–43, 109
Electron neutrino, 55, 56
Electron waves, 108–109
Electrons, 42–43, 47, 66, 73, 75, 78, 81; in chemical bond, 71, 72; kinds of, 76
Elements, 75; higher, 306–307
Elston, Richard, 378
Energia (rocket), 251
Energy, 103, 125–26; in Earth's interior, 211; in stars, 332; of sunlight, 139, 140, 141, 249–52
Energy sources, 28–30, 189–90; fusion as, 87; in space, 254–55; Sun as, 249–52
Environmental Protection Agency (EPA), 101
Enzyme molecule, 96–98
Enzymes, 73, 105–106, 149–50
Eocytes, 135, 136
Eric the Red, 195
Eros (asteroid), 262
Escape velocity, 337
Ethane, 254
Eubacteria, 134–35, 136
Eukaryotes, 134, 136
Euler, Leonhard, 123
Evolution, 119–20, 125, 152–53, 154, 168, 307; of green plants, 140–41; human, 9, 16, 374, 375; insect, 164, 165
Ewing, Maurice, 203–204
Experimentation: dangers in, 99–101
Extinctions, 146–48, 219, 247; *see also* Dinosaurs, extinction of

Faults (in Earth's crust), 128–30
Feathers, 151, 152, 154, 158; function of, 155, 156
Fegley, Bruce, 227–28
Fibiger, Johannes A. G., 43
Finsen, Niels R., 43
Fire, 25–27, 28
Fireflies, 102–103
Fish, 31–33, 137–38, 171
Fission, 87
Flight, 152, 155, 157–61
Flowering plants, 165
Fluorescence, 67–68, 102
Follini, Stefania, 114–16
Forests: loss of, 28, 29, 190
Forrest, William, 348
Fossil fuels, 29–30
Fossil record, 140, 203
Fossils, 138, 202; dinosaurs, 142–43, 145; early man, 9, 11, 13, 17–18, 21, 27; fake, 151–53; of intermediate life-forms, 151–53, 154–56
Franzblau, Edward, 176–77
Free radicals, 72–74

Frere, John, 7
Fricke, Hans, 33
Frobisher, Martin, 195
Fuels, 28–30, 87
Fundamental particles, 76–77, 78–80
Fusion, 63, 87, 297; cold, 87–89, 90–91

Galaxies, 314, 320, 335, 373, 376; active centers of, 338–39, 344; arrangement of, 359–60; clusters of, 354; in collision, 349–51; distance from Earth, 340–42, 356–57, 361–62, 365; quasars as, 352–53; spectra of, 358–59
Gamma rays, 314, 336
Ganymede, 253
Gas clouds, 224–25, 238, 239, 314, 320, 321–22, 360
Gases, 57
Gasoline, 252–55
General Electric, 84
General theory of relativity, 259, 260, 322–23, 324, 343; test of, 325–28
Genes, 97, 105–106, 107
Genetic engineering, 99–101
Genetic information, 187
Genome project, 105–107
Gigantopithecus, 161
Giotto (rocket probe), 225
Glaciers, 202, 203
Glaser, Peter E., 251
Globular clusters, 338, 341–42
God, 374
Golden, Susan S., 141–42
Goldsmith, Roger, 198–99
Goliath beetle, 158
Gorillas, 9, 14, 19, 161
Gould, Stephen Jay, 167
Grand Unified Theories, 244–46
Graphite, 239, 240
Gravitational field, 257, 370
Gravitational interaction, 54, 79
Gravitational lens, 343, 344–45, 353–54
Gravitational pull, 322, 347, 350, 364; of pulsars, 368, 369; sources of, 256–57
Gravitational waves, 259–60, 322–25
Graviton, 315
Gravity, 297, 298, 337–39, 343–45; weakness of, 323
Great apes, 19
"Great Winter," 203, 204, 205
Green plants, 139–42
Green turtles, 166–69
Greenhouse effect, 180, 187–90, 205, 300, 301
Greenland, 194–96
Grimaldi, David, 163
Grimm brothers, 24
Guanahani (island), 197–99

Half-life (neutrons), 317, 319
Halley, Edmund, 248
Halley's Comet, 217–19, 220–22, 223, 225, 270, 271, 289
Halobacteria, 135, 136
Hands, 14, 16–18
Harris, William, 342
HD 114762 (star), 309
Heat, 130; in center of Earth, 131–33; from Sun, 220, 297, 298
"Height of Up, The" (Asimov), 330
Helium, 45, 46–47, 49, 211, 223, 239, 306, 319; liquid, 46, 49, 52, 60, 61, 94–95; lost, 93–95; in Sun, 297
Hemoglobin, 39–40, 97
Heredity, 140, 141
Herman, Alan, 64–65
Hermes (asteroid), 235
Higgs particle, 53
"High Energy Astrophysics Observatory 3" (satellite), 336
Himalayan Mountains, 144, 204, 205, 212
Hoaxes, scientific, 152, 153
Hominids, 8–9, 10–12, 13–15, 16–18; discovery of use of fire, 25–27; speech, 19–21
Homo (genus), 9, 11, 17
Homo erectus, 17, 20, 26–27

Homo habilis, 9, 17, 18, 20
Homo neanderthalensis, 11
Homo sapiens, 8, 11, 17, 25, 26, 27, 163, 222
Homo sapiens neanderthalensis, 11, 20
Homo sapiens sapiens, 10, 11, 12, 19
Hot spots, 120–21, 125–27
Hot springs, 131, 135, 136
Hough, Jim, 324
Hoyle, Fred, 152, 153
Hubble, Edwin P., 361–62
Huebner, Walter E., 225
Humanity, 7–9, 74, 372; age of, 7, 9, 10–12, 34, 220, 222; and existence of universe, 373–75
Hydrocarbon molecules, 229–30
Hydrogen, 93, 94, 211, 223, 239, 306, 319; critical temperature/pressure of, 58; forms of, 90, 91; as fuel, 29–30, 190; liquid, 49, 50; in space, 255; in Sun, 297
Hydrogen atoms, 28, 29, 30, 72, 173; in benzene, 69–71
Hydrogen bombs, 90–92
Hydrogen fusion, 90–91; in stars, 346, 347
Hyoid bone, 20–21

Iapetus, 263
IBM, 71
Ice ages(s), 121, 202–205, 247
Impact theory (mass extinctions), 146–48, 178–79, 180
Incandescent light, 102
India: collision with Asia, 121, 144, 204–205, 212
Indo-European family of languages, 23–24
Industrial Revolution, 28
Insects, 138, 158, 162; as successful life-form, 163–65
Intelligent life, 302, 310
Io, 213
Iodine monocyanide, 67–68
Iridium, 145, 216
Iridium layer, 178
Iron, 132–33, 232
Iron core (Earth), 131–33
Island chains, 121, 143
Isovaline, 149

Jones, J., 220
Jones, Sir William, 23
Jupiter, 58, 95, 211, 213, 230, 256, 257, 279, 309, 346; angular momentum, 272; as black dwarf, 347; rotation, 367; satellites of, 253; space probes to, 256, 258

Kekulé, Friedrich A., 69–71
Kelvin, Lord (William Thomson), 124
Kiran, Erdogan, 58
Koch, W., 45, 47
Kori bustard, 158, 160
Kowal, Charles, 288–89
Krypton, 45, 46
Kuiper, G. P., 253
Kutzbach, John E., 202, 204–205

Lake, James A., 135–36
Land life, 136–38, 183, 184
Land temperature, 200
Land vertebrates, 31, 32
Landmasses: collision of, 121
Language(s), 22–25
Lanthanum, 50, 64
Large Electron Positron (LEP), collider, 79–80
Large Magellanic Cloud, 307, 316, 321–22, 328, 349, 365, 367
Lartet, Édouard, 7
Larynx, 19–21
Laser beams, 124
Latham, David W., 309
Latimer, Miss, 32
Laue, Max von, 39
Lawrence, Ernest O., 51–52
Laws of nature, 373, 374–75
Leakey, Louis, 14
Leakey, Mary, 14
Leakey, Richard, 14
Leap second, 191–93, 283
Lederman, Leon, 54–56

Lehn, Jean-Marie, 96
Lemaître, Georges, 361
Leptons, 76, 77, 78, 79, 80
Life, 247, 374, 375; lightning and, 175–77; on Mars, 285–87, 300, 301; origin of, 125–27, 217, 223–25, 233; possibility of extraterrestrial, 299–302, 310
Life-forms, 183; successful, 163–65
Light waves, 42, 108–109; of quasars, 377–78
Lightning: and life, 175–77
Light-years, 355–57
Lippmann, Gabriel J., 43
Liquid-nitrogen temperatures, 60, 61
Liquids, 57, 58
Lithium, 322
Livermore, Roy, 120–21
Lizard-bird, 151–53, 154–56
Local Group (galaxies), 349–50
Luciferase, 103, 104
Luciferin, 103, 104
Luminosity, 333, 340–42
Lynds, C. Roger, 344–45

McHone, John F., 147
Magellanic Clouds, 340
Magnetic field, 83, 109, 259
Magnetic north pole, 120
Magnets, 48–49, 51; superconducting, 60–61, 62
Mammals, 32, 161, 169–71
Man, modern, 17, 19, 20, 34; ancestors of, 7–9, 10–12; *see also* Hominids; and entries beginning with *Homo*
Mariner 2, 259
Mars, 212–13, 230, 235, 289, 314; possibility of life on, 285–87, 300, 301; manned flight to, 294, 296; Phobos, 282–84
Marsupials, 170
Mass: mystery of missing, 345, 346–48, 360
Mass extinctions, 146–48
Matter, 35, 316, 322–23; and antimatter, 314; gravitational pull of, 337; swallowed by black holes, 338, 339
Meech, Karen J., 290
Mercury (element), 48
Mercury (planet), 212, 230, 274–75, 299
Mesons, 76
Metal(s), 26, 232, 242
Meteor swarms, 221–22
Meteorites, 41, 147, 149, 232–34, 269; ages of, 8; amino acids in, 150; diamonds in, 238–40; iridium in, 178; from Mars, 285–87; in Siberian explosion, 214–15, 216
Methane, 85, 125–26, 223, 229, 261; on Pluto, 278; on Titan, 253–54
Methanogens, 135, 136
Meylan, Anne B., 167
Microorganisms: genetically altered, 99–101
Microscopes, 42, 108
Microwaves, 229–31, 242, 250, 251
Mid-Atlantic Ridge, 167
Middle East, 232
Milankovich, Milutin, 202–203
Milky Way, 321, 349–50, 352, 354, 376; globular clusters in, 341; missing mass in, 360
Miller, S. I., 127
Millipede, 136, 139
Minerals, 127, 242
Mitochondria, 140, 141, 167–68
Moas, 162
Molecules, 39–41, 66; atom connections in, 69; and free radicals, 72–73; largest, 223–25; radio waves of, 224
Monotremes, 170–71
Monsters, 160–63
Moon (Earth), 212, 246, 252–53, 255, 299, 314; base on (proposed), 295; craters, 207–208; detecting changes in position of, 124; formation of, 207; gravitational effect of, 192, 206, 235; and history of

Moon (*cont'd.*)
Earth, 205–208; man's landing on, 205–206; orbit of, 282–83; resources of, 241–43
Moon rocks, 207, 286
Morden, Clifford W., 141–42
Morrison, Leslie V., 248, 249
Moseley, Henry G. J., 44
Mountain chains, 204–205
Mueller, K. A., 63–64
Muon neutrino, 55–56
Muons, 55, 56, 78, 88–89
Mutual annihilation, 82

NASA, 251
Natal homing, 166–69
Natural gas, 28–29, 30, 94, 189–90, 253
Natural History Museum (London), 153
Neanderthal man, 10–12, 17, 20–21
Neodymium, 62, 363
Neon, 45, 46, 49
Neptune, 211, 254, 256, 258, 288, 370; discovery of, 255; orbit of, 257; space probes to, 260–61, 263, 311
Nereid, 261–63
Neutrino beam, 54–56
Neutrinos, 54–56, 76, 78, 245–46, 315–16, 327–28; detection of, 327, 329; from supernovas, 328–31
Neutron star(s), 330, 332, 334–35, 336, 367
Neutrons, 75–76, 78, 81, 240; half-life of, 316–19
NGC 5548 (galaxy), 339
Nickel-iron meteorites, 232–34
Nitrates, 175–76
Nitric acid, 176
Nitrogen, 94, 175–76, 229, 253–54, 255, 261; liquid, 50, 60, 61
Nitrogen atoms, 29, 265
Nitrogen dioxide, 176–77
Nitrogen-fixing bacteria, 176
Nitrogen oxides, 180
Nobel, Alfred B., 43
Nobel Prize, 42–44, 54, 56, 63, 79, 81, 82, 96, 105, 110, 187
Noble gases, 45–46
Nordihydroguaiaretic acid (NDGA), 73–74
North America, 119, 120, 144; discovery of, 194–96; movement of, 204–205
North Pole, 122, 196, 200, 203, 204; position of, 119, 120, 121
Novas, 305
Nuclear bombs, 216, 326
Nuclear energy, 190
Nuclear ignition, 279, 281, 320
Nuclear magnetic resonance, 147
Nuclear waste storage, 172–74
Nucleic acids, 59, 105–106, 107, 127, 135, 175, 223; in green turtles, 167–68; structure of, 39
Nucleotides, 106, 135
Nutation (of Earth), 123

Ocean(s), 143, 211, 212, 213
Oceanic climate, 200
Oceanic hot spots, 125–27
Oil, 28–29, 30, 189, 190
Oil spills, 184, 186
Onnes, Heike K., 48
Oort, Jan Hendrik, 270
Oort cloud, 271–72, 371–72
Organelles, 140–41
Organic compounds, 233, 286, 287
Orion Nebula, 320
Ornithischian dinosaur, 142
Ornithorhynchus, 170–71
Ostrich, 162
Overkill, 178–80
Oxygen, 46, 50, 62, 184–85; discharged by plants, 139–40; ozone as form of, 182–83
Oxygen atom, 28, 47, 173
Oxyluciferin, 103
Ozone layer, 125, 137, 251; hole in, 181–83

Pacific plate, 121, 128–29
Pangaea, 143–44, 201–202
Parallax, 312, 313

Particle accelerators, 51–53, 60–62, 82
Particles, 76–77, 78–80; opposites of, 81–82
Past (the): seen when looking at distant objects, 352–54, 355–57, 377
Pauling, Linus, 71
Peary, Robert E., 196
Pedersen, Charles J., 96
Pelican, 160, 161
Penguins, 31, 184, 185
Petrosian, Vabe, 344–45
Phobos, 282, 283–84
Photons, 315
Photosynthesis, 30, 188
Physics, 2
Pinhead: inscribing letters on, 111–13
Pioneer 10, 256–57, 258–60
Pioneer 11, 256
Pions, 55
PKS 1145-071 (quasar), 354
Planetesimals, 262
Planets, 211, 255–57, 279–80; angular momentum, 272; formation of, 218, 262, 299–300; habitable, 301–302; life-bearing, 299–300; major/minor, 274; search for, 308–10; size range of, 274–75
Plant tissue, 139–40, 265, 266
Plants, 138, 139, 144, 165; green, 139–42
Plate movement, 119–21, 143–45, 201, 212–13; on Venus, 213
Plate tectonics, 143–45, 213
Plateau effect, 202–205
Pluto, 256, 258, 263, 288, 371; Charon, 276–78; planet or asteroid, 273–75
Pollution, 35, 186; in space, 291–93
Polymers, 225
Popp, Carl, 176–77
Population rise, 34–36, 74
Positron(s), 80, 81
Posture, upright, 13–15, 16
Potassium, 8–9
Precession of the equinoxes, 122
Prinn, Ronald, 227–28
Prochlorothrix, 141–42
Prokaryotes, 134, 135, 136
Propane, 254
Protein molecules, 96–97
Proteins, 39, 73, 127, 175, 223; amino acids in, 148, 149
Proton beam(s), 54–55
Protons, 59, 75–76, 77, 78, 81–82, 317, 318–19; breakdown of, 244–46; in hydrogen, 90
Proxima Centauri, 371, 372
Pseudodontron, 160
Pteranodon, 157
Pterosaurs, 157–60, 161
Pulsars, 367–69
Pygostyle, 156

Quantum mechanics, 47
Quantum theory, 44, 374
Quarks, 53, 75–77, 78–79, 80
Quasars, 320, 343–44, 351, 352–54; red shifts, 377–78

Radar, 213
Radiation, 72, 203, 245, 279–80, 320, 348; from galaxies, 335, 350–51; of quasars, 378
Radical(s), 72
Radio telescopes, 224, 305, 353, 357
Radio waves, 320
Radioactive dating, 8
Radioactive elements, 331–32, 333
Radioactive wastes, 92, 190
Radioactivity, 8, 91, 92, 363
Radon, 45, 46
Rare earth elements, 64
Red dwarfs, 346–47
Red giants, 239–40, 297, 298, 332
Red shift, 376–78
Redwoods, 162
Relativity, 44; *see also* General theory of relativity
Relativity test, 325–28
Reptiles, 32, 152, 159; winged, 157–61

Rhinoceros, 161
Ribosomes, 135, 140, 141
Richardson, Philip, 198–99
Richie, John P., Jr., 73, 74
Rock formations, 206
Rocket probes. *See* Space probes
Rockets, 226, 227, 291, 300
Rocks: Moon, 207, 286; from space, 232–34
Rocky mantle (Earth), 131–32
Rocky Mountains, 204, 205
Rous, Francis P., 44
Rubbia, Carlo, 79
Ruddiman, William P., 202
Ruska, Ernest, 42–43, 109
Rutherford Laboratory, 89

Salmeron, Miguel B., 109, 110
Salt mines, 172–74
Samana Cay, 199
San Andreas fault, 128–29, 130, 212
San Francisco earthquake, 128
San Salvador, 198–99
Sanduleak (star), 365
Satellites, 211, 252–53, 274; artificial, 124, 291, 292–93
Saturn, 211, 253, 261, 263, 288
Scanning electron microscope, 109–10
Scanning tunneling microscope, 71
Schaefer, Bradley E., 261, 262
Schaefer, Martha W., 261, 262
Schevoroshkia, Vitaly, 24
Schliemann, Heinrich, 197
Schwartz, Melvin, 54–56
Sciama, E. W., 375
Science, modern, 1–3
Sea level, 188–89
Sea life, 136–37, 184–85
Seabirds, 160–63
Seafloor spreading, 167
Seals, 31, 184, 185
Segrè, Emilio, 82
Sequoia trees, 162
Seyfert galaxies, 338–39
Shelton, Theophilus, 248–49
Short-period comet, 270
Siberia, 200–201, 204, 236–37; blast over, 214–16
Signal-detecting instruments, 231
Silicon dioxide, 146, 147
Sillen, A., 27
Sirius, 356
Smith, J. L. B., 32
Soil bacteria, 181, 183
Solar cells, 250–51
Solar energy, 190, 242
Solar flares, 259, 264–66
Solar Maximum Mission, 269
Solar power stations, 249–52
Solar system, 206, 211, 217, 241, 269, 270, 308, 314; age of, 8, 233, 238; angular momentum in, 272; comets in, 217–18, 219, 220–21; formation of, 224–25, 238–39, 307, 320; kinds of bodies in, 273–74; life in, 300; Oort cloud in, 271–72; outer, 289; satellites (natural) in, 252–53; travel in, 296
Solar wind, 220, 239, 259, 264, 265, 270, 370–71
South Pole, 200
Soviet Union, 52–53, 85–86, 251–52; space exploration, 241, 243, 292, 294
Space exploration, 293–96; cooperation in, 243, 250, 252, 294
Space flight, 293; manned, 294–96
Space junk, 251, 291–93
Space probes, 222, 223, 225, 229, 230, 253, 255, 256–57, 300, 311–13; Halley's comet, 271; *Pioneer 10*, 258–60; *Viking*, 285
Space settlements, 35, 296–97
Space Shuttle, 293
Space station(s), 294–95
Space travel, 299
Spectrum(a), 358–59, 360, 376, 378
Speech, 19–21
Speed of light, 68, 325–28, 337
Spinrad, Hyron, 357

Spiny anteaters, 170
Springtails, 164
Standard human genome, 106–107
Star clusters, 365–66
Star formation, 224, 279, 308, 348, 373
Stars, 192, 239, 260, 302, 314, 320; age of, 331–32, 363; birth of, 320; brightness, 340–41, 342, 366; chemistry of, 362–63; collapse of, 323, 334; core temperature of, 364; distant, 312–13; events at center of, 306, 329–30; exploding, 327; mass of, 364–66; motion of, 122; seeing in the past, 355–57; sizes of, 346–48; spectra of, 360; wobbling, 309–10
Steinberger, Jack, 54–56
Stellar winds, 239, 240
Stishov, S. M., 146
Stishovite, 146–48
Stratosphere, 182
Strobel, Gary, 100–101
Strong interaction, 79
Subatomic particles, 108, 317, 326
Subatomic physics, 53, 61
Sulfur, 133, 135, 136
Sulfuric acid, 177, 227–28
Sun, 95, 230, 273–74, 308, 339; angular momentum, 272; charged particles from, 258–59; companion star to, 257; death of, 297–99; as energy source, 249–52; fusion in, 88; internal heat of, 364; mass of, 346; objects striking, 267, 269; reliability of, 246–49; wobbly path of, 309
Sun-grazers, 267–69
Supercollider, 60–62
Superconductive materials, 41, 62, 113
Superconductivity, 41, 48–50, 95, 113; competition for research funds for, 60–62; thallium and, 63–66
Supercritical carbon dioxide, 59
Supercritical fluids, 57–59
Supercritical water, 58, 59
Supernova 1987A, 367–69
Supernovas, 239, 240, 265–66, 305–307, 319, 349, 360; brightness of, 341; light from, 321, 327; neutrinos from, 316, 328–31
Superstars, 363–66

TAU project, 312–13
Tauon, 78
Tauon neutrino, 56
Taurus (constellation), 348
Technetium, 48
Technology (human), 26, 28
Tectonic plates, 203–205, 212–13
Telescopes, 123, 305, 353, 357, 365; fastest, 358–60; "gravitational," 324–25
Temperature, 185, 187, 188–89; at center of Earth, 131–33; in fusion, 88; in superconductivity, 48–51, 52, 61, 63–65
Temperature cycle(s), 203
Tempest, Will, 248, 249
Thallium, 62–65
Thermoluminescence, 12
Thorium, 46, 93–94, 363
Tidal effect, 277, 282–83, 284
Tidal waves, 236
Tides, 191–92
Time: and distance, 352–54, 355–57, 377
Titan, 229–31, 253–55, 261, 263
Toolmaking, 12, 16, 17–18
Tools, 7, 9
Top-quark, 78–79, 80
Tree-ring calendar, 265–66
Tritium, 89, 90–92
Triton, 254, 255, 261, 263
Tsunami, 179
Tunguska incident, 214–16
Tunneling effect, 109
Turner, Sean, 142
Two-neutrino experiment, 55–56

"Ugly Little Boy, The" (Asimov), 21
Ultraviolet light, 72, 102, 125, 127, 137; and ozone layer, 181, 182
Undulator, 41
United States, 53, 182; space program, 241, 292, 294
Universal recession, 361
Universe: age of, 2, 83, 331–34, 340, 361–63, 372; basic objects making up, 75–77; beginning of, 53, 153, 306, 318–19, 356–57, 378; chemistry of, 360; end of, 376–78; expansion of, 345, 359; expansion rate of, 331, 333, 340; formation of, 314; human beings and existence of, 373–75; multiplicity of, 374–75; size of, 365–66; total matter in, 35
Up-quarks, 78–79
Uranium, 8, 46, 93–94
Uranus, 211, 256, 260, 261, 288–89; discovery of, 255; orbit of, 257; space probes to, 311

Van Biesbroek 8 (VB 8), 280–81
Van Biesbroek 8B (VB 8B), 280–81
Venus, 213, 226–28, 235, 259, 299; chance of life on, 300; microwave echoes from, 230, 231
Vertebrates, 31, 32, 138
Viking probes, 285, 286, 287, 314
Vikings, 194–96
Vinland map, 193–96
Voyager 2, 229, 253, 254, 260–61, 263, 311, 370–72
Volcanism, 125, 145–48, 180; on Venus, 227–28
Volcanoes, 131, 132, 143, 146, 212–13; on Venus, 226

Walking erect, 13–15, 16
Water, 30, 137, 184–85; heat capacity of, 199–200; old, 172–74; *see also* Supercritical water
Water molecules, 173
Watling, John, 198
Watling's Island, 198
Watson, James Dewey, 105, 106, 107, 110
Weak interaction, 54, 55, 79
Weapons, 12
Weber, Joseph, 323–24
Whales, 31, 185
Whipple, Fred, 218
White dwarfs, 299, 331–34
Wildfires, 178, 179–80
Winds, 124, 204–205
Winget, Donald E., 333
Wright, Ian P., 286, 287

X-ray diffraction patterns, 39–40, 110, 147
X rays, 42–43, 72, 108–109, 110, 113; and black holes, 335–36, 338; ultrapowerful, 39–41
Xenon, 240, 311, 313

Yun, Joao L., 328

Z degree particle, 79–80
Zewail, Ahmed, 67